The Patrick Moore Practical Astronomy Series

For further volumes:
http://www.springer.com/series/3192

Grab 'n' Go Astronomy

Neil English

 Springer

Neil English
Fintry by Glasgow
UK

ISSN 1431-9756   ISSN 2197-6562 (electronic)
ISBN 978-1-4939-0825-7   ISBN 978-1-4939-0826-4 (eBook)
DOI 10.1007/978-1-4939-0826-4
Springer New York Heidelberg Dordrecht London

Library of Congress Control Number: 2014937898

Cover illustration: Cover image used with permission from Astronomics.

Printed on acid-free paper

Springer is part of Springer Science+Business Media (www.springer.com)

Preface

Amateur astronomers lead busy lives. After a long day at the office, or looking after young children, the last thing you'd like to do is to drag out a large telescope and have to wait for more than an hour to receive adequate gratification at the eyepiece. This niche is often better suited to a small, grab 'n' go telescope and/or binoculars. In addition, more and more of us are on the move, traveling for both business and pleasure, desiring to bring a small, portable instrument along for the ride.

Or maybe, you're one of the growing army of casual observers, not quite committed enough to own a larger instrument but curious enough to see what a smaller, portable setup can accomplish. Perhaps you live in a high-rise apartment building with limited room to use a large telescope. Is astronomy an anathema? Are you a daytime astronomer, interested only in solar observing? Maybe you're a dedicated lunar and planetary observer, or do you enjoy seeking out the treasures of the deep sky? If your observing schedule fits one or more of the above scenarios, then you need an observing book dedicated to your fastidious needs—enter grab 'n' go astronomy.

After surveying the rich, varied, and constantly changing milieu of grab 'n' go telescopic culture (Part I) and their accessories, this book sets out to demonstrate the amazing things that can be achieved, even with a modest investment.

After surveying the market for small, ultra-portable telescopes, dedicated chapters on solar, lunar, and planetary are presented before delving into the rich milieu of deep sky objects on offer to both suburban and rural observers. The latter will be divided into separate chapters, where we present the full pantheon of celestial objects—including open and globular clusters, emission and planetary nebulae, galaxies, double and variable stars—on show at different times of the year.

The observing part of the book (Part II) is designed to get the observer to think about the objects he or she is viewing. Quite a bit of science is included in the text to help you decipher what you're seeing. In addition, questions have been included to ponder, as well as activities to share in and moments to just pause for thought. This approach resonates well with my conviction that grab 'n' go astronomy can be educational as well as fun to undertake.

Each chapter can be read independently of the other, and the reader is encouraged to dip in and out of the book as and when appropriate. Above all, we hope that this will encourage folk to maintain their interest in observational astronomy and disseminate that knowledge to young people and novice adults alike.

Fintry, UK   Neil English
April 2014

Acknowledgments

I would like to thank Alexander Kupco from Ricany, Czech Republic, for allowing me to use many of his fine visual illustrations recorded with his 80 mm Zeiss AS refractor. My thanks also to Stanislas Maksymowicz for permitting me to use his drawings of Venus conducted in daylight. Gratitude is extended to Mike Pearson, an avid astrophotographer from Glasgow, for allowing me to make use of some of his images. I would also like to thank John Watson for turning an idea into a working proposal and the hard working team at Springer, especially Maury Solomon and Nora Rawn, for their expert assistance in the processing of the manuscript. Last but not least, I would like to thank my wife, Lorna, and sons, Oscar and Douglas, for putting up with my long absences from their presence in preparing the material used in the book. Thank you for your love and patience.

Contents

About the Author

Dr. Neil English has a BSc in Physics and Astronomy, and also a PhD in Biochemistry. He is a Fellow of the Royal Astronomical Society and a regular contributor to *Astronomy Now* (the UK's major astronomy magazine), as well as to Ireland's *Astronomy & Space*. Neil's astronomical images have been published in various magazines and journals, including a full page in the June 2006 issue of *Astronomy*. Neil has, and continues to, made contributions to *Cloudy Nights* in the form of several detailed telescope reviews. He is the author of *Guide to Mars* (Pole Star Publications, 2003) and he has a number of books published by Springer, including *Classic Telescopes, Choosing, and Using a Refracting Telescope* and *Choosing and Using a Dobsonian Telescope*. Neil currently has (among other instruments) a large 12-in. Dobsonian, which he uses on the darkest, steadiest nights at his home in rural Scotland.

Part I

The Equipment

Chapter 1

Grab 'n' Go
Binoculars

Anyone with even a casual interest in astronomy should own a good pair of binoculars. Even avid amateurs with a houseful of telescopes will have at least one pair of binoculars at the ready. And for some activities, binoculars will be all you really need for a lifetime of applications.

It is not unusual for amateur astronomers to begin their work not with a telescope but with binoculars. You'd be amazed at just how much those simple optical accoutrements can extend naked-eye vision. Large open clusters such as the Pleiades look magnificent through them, as does the expansive Andromeda Galaxy (M31) on an autumn evening. The decent light-gathering power and magnification can pull out many thousands of stars in the summer Milky Way that could not be seen otherwise. And watching the changing aspects of the Moon's phases is simply spellbinding.

If your main interest in astronomy is exploring the fine details on planets or showing structure in distant galaxies, you will eventually want to get a telescope, as binoculars just don't have enough magnification (and light-gathering power) to do justice to these objects. However, binoculars certainly have their advantages over telescopes for astronomy, and an ultra wide field of view of them is tremendously desirable. If you're a casual deep sky observer, determined to learn the constellations, or a dedicated comet hunter or variable star observer (discussed later in the book), then good binoculars may be all you ever need.

Choosing Binoculars for Astronomy

There is a bewildering variety of binoculars available on today's market, and it certainly isn't easy to recommend one model over another. The best binoculars for daylight applications may not be the best for nighttime use. Small binoculars are

N. English, *Grab 'n' Go Astronomy*, The Patrick Moore Practical Astronomy Series, DOI 10.1007/978-1-4939-0826-4_1, © Springer Science+Business Media New York 2014

Fig. 1.1 Binoculars are indispensible tools for the grab 'n' go astronomer (Image by the author)

easier to carry about and can be held more steadily in the hand, but larger models can gather more light and allow you to see more details in the celestial objects.

Before looking at a few specific models, let's first discuss the parameters you need to evaluate before making your choice.

Aperture

This is usually provided by the large number printed on the binocular. The larger the aperture the greater the amount of light gathered and the fainter the objects seen. In general, 30 mm is considered about the minimum, but some binocular astronomy enthusiasts employ units with clear apertures as large as 127 mm.

Magnification

The ideal magnification on a set of binoculars to be used for astronomy will differ depending on the way you intend to use it. Most astronomy binoculars have magnifications of between 7 and 15. The higher the magnification, the heavier the binocular will be. Lower powers, although providing wider fields of view are not the best at seeing fainter details. For example, a typical 7×50 binocular will be less effective at seeing fainter stars than a model having 10×50 specifications.

Hand-Held or Tripod Mounted?

Hand holding a binocular is an enjoyable and less expensive way to enjoy astronomy, with many "normal"-sized binoculars being less expensive than the larger "giant" ones, and there are no accessories to buy; there is the added advantage of no setup time. Also remember that mounting binoculars on a tripod and looking through them is not that practical for viewing positions that are far from the zenith.

Smaller binoculars also have the added bonus in that they are far more versatile, and you can use them for many other applications unrelated to astronomy. If you plan to use this method, you should keep magnification below 12× in order to maintain hand-held steadiness. A good pair of binoculars with a magnification of 7 to 12× and a large objective lens will show you the moons of Jupiter, hundreds of star clusters, nebulae and even some galaxies.

Larger binoculars that can deliver magnifications of between 15 and 30× need to be mounted on a tripod. These binoculars will show more detail and resolve more stars, though they still won't turn your binocular into a telescope. However, there is nothing like the view in, say, a 25× 100-mm binocular to take your breath away on a dark, clear night.

If you are going to use your binoculars for astronomy and don't want the hassle of using a tripod, 7 × 50 binoculars have proven to be a perennial favorite among sky gazers. However sometimes you may want to sacrifice some field of view in order to go deeper into space. In this capacity, a 10 × 50 binocular is even more popular. Most people are just able to hand hold a 10 × 50 for a prolonged length of time and keep the image steady enough for a good view, providing you a nice balance between detail seen and deep sky penetration.

Binoculars with objectives of 60 mm or larger are called giant binoculars, but it is possible to hand hold a 10 × 70 or even an 11 × 80 for short periods of time by grasping them around the barrels near the objectives. Larger objective lenses will reveal fainter stars and probably more detail but will become too heavy to hold steadily for any length of time.

The increased magnification of 10 × 70 binoculars, for example, will add a little more detail to what you can see in a 7 × 50 and maintain the same image brightness as well. So, as far as hand-held binoculars are concerned, 10 × 70 or 11 × 80 represent the upper limit of what is practical for hand holding. For extended use, you will have to go with some sort of mounting arrangement.

Binoculars with magnifications over 15× and objectives of 70–80 mm or more can rival and exceed the view of some small telescopes for certain types of objects, and you get the comfort of using two eyes. The one downside to most giant binoculars is their fixed magnification, though a few models are now available that offer interchangeable eyepieces.

Most large, heavy-duty camera or video tripods will work for 80 mm and some 100 mm binoculars. Be sure to compare the weight of the binocular with the maximum load capacity of the tripod, if listed. Unfortunately, this figure is not standardized, nor will it guarantee how well it will work with a large binocular, but it is a place to start.

For 80-mm binoculars, look for a tripod that lists a capacity of at least 12 pounds, but a 15-pound unit will be noticeably better. Another thing to check is the actual weight of the tripod. Light tripods will struggle when loaded with a binocular of equal weight or more. Generally because portability is rarely an issue when it comes to tripods for astronomical binoculars, bigger is always better. Quick-release plates are a convenient feature to look for, but check to make sure they fit very tightly. Loosely fitted binoculars on the tripod head will produce unwanted sag.

Most standard tripods can be used, but because you are looking upwards, it does mean that the eyepieces will be in an awkward position. The best way to get around this is to use a chair and position yourself almost under the tripod. With traditional tripods this can be a little awkward, as the legs often get in the way. Sophisticated binocular mounts, such as the Vanguard Alta Pro, have an adjustable

Fig. 1.2 Binocular quick-release plates (Image by the author)

Fig. 1.3 Binocular mounts can be as simple or as elaborate as you like (Image © John Clancy. Used with permission)

central column that you can use to position your binoculars away from the center of the tripod so you can more easily position yourself under your optics. Recently companies such as Orion Telescopes and Binoculars have introduced very cost effective parallelogram mount for $499.

Exit Pupil

Although the size of the objective lens determines to a large degree how much light can enter the binoculars, it does not completely determine how much light enters your eyes, which is actually a more important consideration. A measurement known as the *exit pupil* gives you the width of the column of light exiting the eyepiece and is calculated by dividing the objective lens size by the magnification of the binoculars.

In dark and poor light conditions, the maximum pupil size of a human eye is typically between 5 and 9 mm for people under 25 years of age (with the mean being about 7 mm). This maximum size will also decrease slowly with age. So apart from the very small benefit of ease of use, there is not much point in owning binoculars with an exit pupil larger than your pupil. That said, an exit pupil smaller than your own eye pupil will result in a darker image.

Here's a couple of examples that illustrate these ideas.

For a pair of 7×50 binoculars in very low light conditions, the pupil diameter is 7 mm, so the exit pupil of binoculars will be $50 \div 7 = 7.1$ mm.

Because the human pupil is about the same size as the binoculars' exit pupil, the emergent light at the eyepiece then fills the eye's pupil, meaning no loss of brightness in low light conditions when using these binoculars (assuming perfect transmission). Thus the result is that you will perceive the image as being as bright as if you were to see it with the naked eye.

Now if you were to use much smaller 8×20 binoculars in very low light conditions, the exit pupil of binoculars would be $20 \div 8 = 2.5$ mm. Because the 2.5-mm exit pupil of the binocular is smaller than the 7-mm human pupil, you will perceive the image as being dark.

Thus, for astronomical applications you are looking for an exit pupil of 5 or more; however, with higher magnifications this is not always possible, as the objective lenses would have to be massive. So, although many giant binoculars have slightly smaller exit pupils than ideal, they are still large enough to provide you with a sufficiently bright image. However, this brightness is also determined by the amount of light transmitted by the binocular to your eye.

Light Transmission

Although the size of the objective lens and the size of the exit pupil are very important, they only tell part of the story when it comes to just how well your binoculars will perform in very low-light conditions. As light travels through a glass lens or prism, some light is lost through absorption and reflection at each air-to-glass surface or inside the prism system itself. The amount of original light available to the observer by the time it exits the eyepiece will vary from as low as 50 % to as high as 97 %.

The precise transmission levels are dependent on a number of factors, including the quality and number of optical glass elements used in the lenses and prisms, the configuration and size of the prisms, how well the collimation of the optical system is, the type and amount of anti-reflection, as well as the quality of the coatings applied to the lenses and prisms.

The term used to describe this percentage of light that is not lost through the optical system is *transmittance*. Good quality binoculars will usually have a transmittance level above 90 %, while lesser quality instruments that use poorer quality glass and coatings will be far lower. All things considered, it's possible for a 10×40 binocular (exit pupil 4 mm) with a high transmittance (90 %) to actually deliver a brighter image than a 7×35 (exit pupil 5 mm) with a lower transmittance (70 %). What makes matters worse is that most manufacturers don't publish their binocular transmittance levels, unless they are sure that it outperforms their competitors! So how can you be sure that the binoculars that you get have a high level of transmittance?

To avoid disappointment, it is always best to ensure that the binoculars you are purchasing employ quality optical glass and prisms as well as anti-reflection coatings. An easy way to see how well anti-reflection coatings work is to point the binocular at some bright street lights. After focusing carefully, examine the image for internal reflections. Many budget units fail this test. Look for those that employ BAK-4 prisms and not the inferior quality BK-7 glass. Most modern binoculars have anti-reflection coatings on all or at least some of their air-to-glass surfaces. What these coatings do is to assist light transmission.

It is important to note how the manufacturer describes their coatings, as they are not all created equal. Ideally you would like to see the term "Fully Multi-Coated" stamped on the binocular, which means that all the air-to-glass surfaces have received multiple layers of antireflection coatings. If you just see "Fully Coated" or "Multi-Coated" it means only some surfaces have coatings, or they only have a single coating. The latter will not perform anywhere near as well as fully multi-coated instruments, all other things being equal.

If you plan on choosing a roof prism binocular then you need to look at the kinds of coatings used to increase their light reflectivity. Roof prisms have a number of advantages over Porro prisms, but they do have one surface that does not exhibit total internal reflection. It is therefore very important for the binoculars' optical performance that the reflectivity of this surface is raised. In most cases, an aluminum mirror coating is used that has a reflectivity between 87 and 93 % or, better still, that a silver mirror coating (reflectivity of 95–98 %) is used. This light transmission of the prism can be improved by using a dielectric coating rather than a metallic mirror coating. A well-designed dielectric coating can provide a reflectivity of more than 99 % across the visible light spectrum. In the less expensive binocular models, an aluminum mirror coating is used, while higher quality units employ silver or dielectric coatings. Roof prism binoculars that lack high reflectivity coatings on the prisms should be avoided.

In the past, a quality giant binocular would be a very expensive purchase, but recently many more manufacturers have been producing astronomy binoculars—many of which are made in China—that has brought the prices down. What is good for us the consumers is that many of the new Chinese optics are now being made to very high optical standards, and while many may not like to admit it, they perform as well as many far more expensive binoculars made in the west. Some popular brands include Oberwerk, which has plenty of nice features, including collimation screws. Other good sources of astronomy binoculars are Celestron, Meade and the excellent Apogee brand. All of these offer fantastic quality for the price and bring giant binoculars within reach of most people's budgets.

In general, you ought be able to find a quality set of 80-mm binoculars for around $100–300, with 100-mm models starting at about $400. High-end models can cost over $1,000 and come with interchangeable eyepieces and pier mounting options. The best giant binoculars for astronomy are not cheap and are usually made in Japan or Germany. Arguably the finest astronomy binoculars are manufactured by Fujinon, which produces everything from a 70-mm binocular to pier-mounted telescopes. Kowa and Nikon produce great models, too.

Close Focusing Range

Although this is certainly important for daytime observations, it is of little importance when it comes to astronomy.

Waterproofing

There is not much of a need for an astronomy binocular to be waterproof. Only reasonable water resistance is enough, as using them at night can expose a binocular to dew and moisture, which can cause a non-waterproof model to mist up inside the mechanism. In general, better quality binoculars tend to be sealed (nitrogen purged) and fully waterproof as well as fog proof, and so this is one indicator to look out for if you want to make sure that the binoculars you are getting are of a good quality.

Field of View

Although the field of view on astronomy binoculars is not as important as it is for people using their binoculars to view unpredictable, fast-moving objects such as wild animals, it is still fairly important. Field of view is basically the width of the scene that is in view when you look through your binoculars. A wide field of view will make it easier to scan the night sky and find objects when looking through the binoculars.

Everyone talks about magnifications, and there is no doubt that high magnifications yield beautiful views of the Moon, planets and fine detail in some deep sky objects. However, many objects in the sky are too large to fit into the field of view of a high power eyepiece. These objects demand a wide field of view to appreciate their beautiful and delicate morphologies. The other caveat with 'wide field' binoculars is that they may exhibit distorted or out-of-focus star images at the edges of the field. The best advice, as ever, is to try before you buy.

Eye Relief

This is the distance behind the ocular lenses where the image is in focus. So if you wear glasses, you can't get your eyes as close to the lenses, and thus you may need a longer eye relief to project the image beyond the ocular lens on the binoculars.

If you wear eyeglasses, you should be looking for an eye relief of around 15 mm or more, to see the full image. The downside to long eye relief is that it usually reduces the field of view.

Eye cups can affect the eye relief also, as they increase the distance from the oculars to our eyes, but they also help keep stray light away from your eyes while using binoculars. Many eye cups are made from rubber and can roll up or down depending on whether you wear eyeglasses or not. The problem with these is that the constant rolling causes the eye cups to break. Another eye cup design uses a sliding cup, but this can be hard to keep in place. A third type employ eye cups that twist up and down, so they can be left at any position (all the way up to all the way down); some even have click stops at regular intervals with the eye relief distance for each stop marked on the cup so you can get the perfect eye relief for your vision.

Collimation Tests

Thus far, we have tackled the various factors that need be in place before binoculars work optimally, including the types of prisms employed, the effect of anti-reflection coatings and so on. But if the various optical components are not aligned properly— that is *mis*collimation—the user will not be able to use the image and/or field of view will be compromised. Fortunately, there is an easy way to test if the binoculars are miscollimated. Simply hold them up to a bright sky background and examine the shape of the light cone emerging from the eye lenses. It ought to be completely circular. Any departure from circularity is a sure sign that one or more of the components has become misaligned. This is a delicate procedure to fix and so is best performed by a specialist.

Image-stabilized (IS) binoculars have been around for over a decade now. Many who use them report an entirely different viewing experience. The image is magnified without jitters, allowing you to relax and enjoy the view rather than having to 'interpolate' the detail. A good pair of IS binoculars will stabilize both high frequency jitters as well as longer period oscillations caused by body movement or platform shake. All this leads to an enhanced experience, almost as if the optical instrument isn't there.

The benefit is much easier to experience then describe. For example, think about being at an airport, watching an airplane take off. As the plane picks up speed, IS binoculars prove indispensible for maintaining sharpness of the image as it shifts in position. Indeed, the view would be utterly impossible without IS. Pick any resolution target, and the amount of detail that can be pulled from the picture is magnitudes ahead of the hand-held view. Even mounted binoculars fall short in comparison. Increased steadiness also makes it easier to resolve finer and fainter detail.

Fig. 1.4 Canon 18×50 IS binoculars (Image © Canon. Used with permission)

So, what are the drawbacks to IS technology? The price is one legitimate objection. That said, prices on many IS are now comparable to mid-priced non-IS units. Many of the top-end manufacturers now make and market a range of IS models. Canon, Fujinon, Nikon, and Bushnell have also done a good job in controlling prices to bring affordable models to the marketplace. Another disadvantage is that IS units are a bit heavier than traditional binoculars, but not excessively so. Certainly the portability is still there. Like all electronic devices, batteries may be required, but typically they don't add much weight, and the cost for a good set of 4 AA rechargeable batteries is not going to kill anyone's budget. Indeed, most units can be used continuously for hours before running out of power. The biggest drawback to an IS unit is its durability, or rather its lack thereof. Its moving parts—gyroscopes and sensors—may not be the best choice if you are going to be expecting hard knocks during rough use. That said, many users are impressed at just how durable IS binoculars are, given the complexity of their designs.

This brief overview of the binocular market should help you get the right kind for your observational program. Good luck with your purchase.

Chapter 2

Grab 'n' Go
Refractors

Small refractors—those telescopes that use lenses instead of mirrors—have long been favored as the ideal accompaniment for astronomy or nature studies on the move. Small, lightweight and capable of delivering both low and high power views more or less out of the box, it is no small wonder that a rich array of instruments can now be had, with price tags that range from bargain basement to outrageously expensive.

Refractors come in two distinct flavors: achromatic and apochromatic. Achromatic telescopes use tried and trusted crown and flint glass to bring two colors of light to a sharp focus. These are perfectly adequate for most applications—either terrestrial or astronomical—and are less expensive than their apochromatic cousins, which employ special, low dispersion (ED) glass to focus three wavelengths of light. Apochromats give brighter images and throw up less secondary spectrum (violet fringing round bright objects than their achromatic counterparts. Because of their modern glass prescriptions, apochromats command a heftier price tag.

Good achromatic optics are *much* closer in performance to apochromats than is commonly believed. Indeed, it was with achromatic optics that all the splendors of the heavens were unveiled to the observers of yesteryear. Moreover, achromatic images are qualitatively different to apochromats; the image is dimmer at a given magnification than the apochromatic counterpart and may suit some observers more than others. Indeed, this author has vociferously expressed a preference for achromatic optics and helped raise awareness about the rich astronomical culture they once commanded.

Up until fairly recently, there was only a limited range of spotter-type 'scopes available for consideration. These were rather fastidious in their design, often

N. English, *Grab 'n' Go Astronomy*, The Patrick Moore Practical Astronomy Series, DOI 10.1007/978-1-4939-0826-4_2, © Springer Science+Business Media New York 2014

displaying fixed magnification ranges—frequently only ranging from low to medium powers—and built with an eye to sating the needs of birders and other wildlife enthusiasts. And although these telescopes can certainly be used to good effect in astronomical projects, they are not ideally suited to the task. Over the last decade, many small refractors have been brought to market that can be used with inter-changeable diagonals and eyepieces and can thus be employed for a broader range of applications that transcend the usual limitations of both spotting 'scopes and dedicated astronomical telescopes.

Another great advantage of using these so-called 'crossover' telescopes is that they can be purchased as so-called optical tube assemblies, and so you can carefully choose the accessories tailored to your needs. This makes them far more versatile than dedicated spotting 'scopes. One can choose either a 1.25-in. diagonal or a 2-in. diagonal, depending on the eyepiece you want to use.

Most birders make do with spotting 'scopes that use relatively lightweight 1.25-in. eyepieces. The 2-in. eyepieces deliver greater fields of view, which is fine for astronomy but normally overkill if you're trying to concentrate on the variegated feathers of a nesting Robin. By purchasing an optical tube assembly, *you* get to choose the kind of viewing you want to experience. Having observed through traditional spotting 'scopes for many years, with their dedicated, non-interchangeable zoom eyepieces, you might find this new-found freedom a great liberation.

Small refractors manufactured primarily for amateur astronomy do not yield correct orientation views but a number of companies, including William Optics, produce both 1.25- and 2-in. prismatic diagonals angled for 45° viewing. They were designed to give very good images over typical daylight magnifications for their small ED 'scopes, like the Zenithstar 66, but the image quality seems to rapidly degrade if powers above 60× or so are employed.

There is a near perfect panacea for this, thankfully, and it is embodied in the form of a high-quality mirror diagonal. In general, its excellent optical flatness and high reflectivity allows you to use much higher magnifications—if your project calls for it—than the 45° prismatic diagonals. The best mirror diagonals have dielectric coatings that boast 99 % reflectivity.

The term 'near perfect palliative' is given for a purpose; the only caveat with mirror diagonals is that, although they yield upright images, the view is reversed left to right. What's more, traditional astronomical mirror diagonals are designed for looking high in the sky and thus are designed with 90° angles. One notable exception is the 1.25-in. Tele Vue 60° Everbrite diagonal. Designed by Al Nagler, this diagonal offers all the comfortable terrestrial viewing of a 45° prismatic diago-nal does but delivers noticeably better images, especially during high-power appli-cations. As you might expect, it doesn't come cheap ($210) either. One birder known to this author uses one with his inexpensive $100 spotter!

Another issue for spotting 'scope users is minimum focus distance, or *back focus*. That's the closest distance to an object that your spotting 'scope will focus on. If you like using your 'scope as a long-distance microscope, you'll need to be able to focus at close range, often within a few meters. If the 'scope you purchase

Fig. 2.1 The Tele Vue 60° Everbrite diagonal (Image © Venturescope. Used with permission)

doesn't come with this information, you'll need to try before you buy. Most commercial spotters can achieve sharp focus at distances ranging from 3 to 6 m. In general the larger the 'scope, the greater the minimum focus distance achieved.

Dedicated spotting 'scopes with non-interchangeable eyepieces tend to be significantly lighter than an equivalent aperture crossover 'scope. Many of the more expensive models are made from ultra-light metal alloys that can be over 50 % lighter than a similar aperture crossover 'scope. The extra weight is not much of an issue when it comes to astronomical applications, when the 'scope is not hauled about as much. Most spotting 'scopes also need to be adequately mounted if a nice steady view is to be enjoyed. We'll be discussing mounting options for these and other 'scopes later in the book. Buying a decent 'scope can be a significant investment. Selecting the right model for your needs and your budget is vitally important. Fortunately, the increased demand for quality optics has led manufacturers to keep their prices down while producing a dazzling assortment of 'scopes from which to choose.

We'll now take a look at one highly regarded traditional spotting 'scope: the Leica Apo-Televid 82. This 3.2-in. (82-mm) aperture spotter is available in a choice of either straight or 45° angled body, but optically these are identical. The focuser can be rotated a full 360° around the mounting collar. On the straight body, this rotation facilitates orienting an attached camera to compose an image as desired. The straight body is a good choice when operating around flat marshy areas devoid

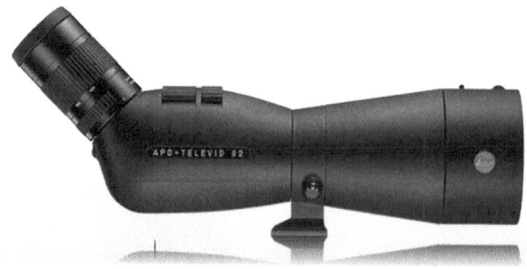

Fig. 2.2 Spotter king: the Leica APO Televid 82 (Image © Ace Cameras. Used with permission)

of trees or high ground. For some people it's easier to sight through the straight body telescope onto a distant target, too.

For quicker sighting when using the camera adapter, or for observing from a car window mount, you should use the straight through view body, too. That said, a straight body will place the eyepiece and spotting telescope at the same height, so a taller tripod or stand will be required compared to that needed for the angled body.

The angled body is the better choice for most users, since this provides more comfortable eye position, particularly when observing objects higher than the observer—a bird perched high in the tree canopy or a craggy cliff edge, or indeed a celestial object. A noteworthy advantage of the angled body, since it too can rotate in the collar, is that the eyepiece can be positioned up, down or to the side to allow people of differing heights to view. When target shooting, the telescope barrel can be rotated so that the eyepiece can be seen from a prone position. For surveillance viewing, the observer can remain hidden around a corner or below a ridge while you are still able to look through the eyepiece. And the view is correct left to right and right side up regardless of the eyepiece position!

The Leica Televid has a special five-element objective (in four groups) containing fluorite for top notch color correction. It has a focal length of 440 mm (so it's about F/5), and the supplied zoom eyepiece delivers magnifications from 25 to 50×. The lenses are coated with an innovative anti-reflection coating called AquaDura that prevents water droplets from adhering, and the surface is also resistant to the formation of fog. Looking forward, we should see similar stuff applied to the objectives of mainstream astronomical refractors. Furthermore, the entire 'scope is nitrogen-filled and water resistant to a depth of 5 m. The 'scope body is made from die-cast magnesium, making the whole package extremely light—just 3.6 pounds (1.5 kg)—for a scope of its size. Being less than a foot (30 cm) long, the Leica Televid 82 is superbly designed for the outdoors in temperatures ranging from 25 to +55 °C.

So, in the cold light of day, is this telescope worth the astronomically high price tag Leica commands for it?

That's a difficult question to answer. It's stylish, rugged, ultra-light and has excellent optics. But it has limited latitude in terms of the range of magnifications it can be used with. Sure, Leica also supplies two very nice additional eyepieces for the Televid 82—a 32× wide-angle and a higher power 20× to 60× zoom—but that's it. Even if it could be charged with higher magnifications, the tiny amounts of color visible at 50× as seen in daylight from a store window would throw up considerably more color under typical astronomical use. Seen in this light the Leica APO Televid 82 is very much a specialist 'scope. After having spent a small fortune for one, a small, premium 'scope really ought be able to do everything superbly! It should be lightweight and compact, have superb, color-free optics and be able to use inter-changeable eyepieces. In short it should perform equally well by day and by night. To that end, in the last few years, an amazing array of small, ultra-portable cross-over 'scopes have made their debut, and they have sold by the thousands across Europe and North America. We'll now take a look at some of these models.

The William Optics Mini-Scopes

The year 2005 provided a bumper crop for small refractor lovers. That's the year William Optics launched not one, not two, but *three* sensational little refractors all at the same time. No doubt it was a bit of an experiment on the part of the company to see which one would win out with the consumer, and at first it was hard to know the winner.

The first to appear was a 66-mm f/6 four-element Petzval, billed as semi-apo. Then came the Zenithstar 66 ED f7 triplet apo, followed fast on its heels by a 66-mm f/6 SD doublet. All came in a beautifully anodized tube complete with rotating Crayford focuser and logoed soft case. All three came with a retractable dew shield and all weigh in at about 5 pounds. As if that weren't enough, all three came with a 1/4–20 L mounting bracket that could be used with nearly any photo-tripod.

The 66 ED Petzval had a nice flat field—a real bonus if you're into photography—but it displayed quite a bit of color in daylight tests at moderate powers (>30×). The single FPL-51 sub-aperture element and Petzval design frankly didn't subdue false color as much as hoped. The triplet Zenithstar 66 was much better in this regard. Daylight and nighttime testing showed only the merest trace of false color, even when pushed to high magnification or bouts of atmospheric thermal instability. The Zenithstar 66 SD then arrived, with an impressive level of color correction. The full aperture FPL-51 element did a superb job at wringing out any chromatic aberration from all but the most stringent testing of objects.

Within a few months of the launch of the Zenithstar 66 f/5.9 SD doublet ($395), the Petzval and triplet models were discontinued. That was probably a wise move on the part of William Optics, as both are more complex and thus harder to manu-facture than the SD doublet. Since 2005, the Zenithstar 66 SD has gone on to become one of the best-selling small telescopes in the world. This little telescope offers excellent optics for its modest price tag ($329 for the optical tube, 1.25-in. diagonal and hard case). Although the latest models sold have black and white

Fig. 2.3 The William Optics Zenithstar 66 SD doublet (Image © Ian King Imaging. Used with permission)

anodized tubes, William Optics cashed in on the incredible popularity of these 'scopes as 'luxury' finders mounted atop larger 'scopes. To this end, they produced both 'Celestron' orange and 'Meade' blue tubes to delight an army of Schmidt Cassegrain fans.

Weighing in at just over 2 kg with a 1.25-in. diagonal and eyepiece inserted, this little 'scope serves up sharp, high contrast and color-free views at low and moderate powers. Even high powers (>100×) reveal only a trace of color fringing around high contrast objects.

Star testing a few of these telescopes showed pretty good results with only minor spherical aberration and a trace of astigmatism detected when pushed to 120× or so. As discussed later, William Optics Zenithstar 66 SD showed a bit of red excess when viewing bright stars at high power, but after reading about Canadian amateur, Clive Gibbons' experiments with short focal ratio ED doublets, the author tried an inexpensive prism diagonal that removed the red excess at the expense of introducing a slight bluish fringe around bright stars and planets at high power.

William Optics also supplies an adaptor that allows the 66 SD to be mated to 2-in. diagonals (in fact it is deliberately designed to take popular SCT accessories) to obtain the widest possible views for a telescope with these specifications. Think about it—a 31-mm Nagler eyepiece would yield a field of view nearly 6.6° wide. That's 13 full Moon diameters!

Nonetheless, despite its appeal as an ultra-rich field 'scope, its aperture (and limiting magnitude of +10.8) restricts its performance as a serious deep sky instrument. Because the 66 SD is so light, it can be easily mounted on a conventional photo-tripod for terrestrial viewing. The dual speed Crayford focuser, fitted as standard on these 'scopes, is a great benefit when homing in on wildlife constantly on the move. The 66-mm doesn't sound like much aperture, but it's enough for most daylight applications using moderate magnifications. The only scenario in which the Zenithstar 66 SD would probably prove lacking is in low light conditions. That said, if you're after an ultra-portable travel 'scope that won't break the bank but nonetheless offers very good, color-free optics, then there is little to go wrong with the William Optics SD doublet.

Since the launch of the Zenithstar 66 back in 2005, the 'scope has been a huge success for William Optics. In the autumn of 2009, the company announced it was ceasing production of this popular model—no doubt a reflection of the global economic recession that preceded it. However, a number of other companies have marketed their own clones of this 'scope, most notably Astronomy Technologies (ASTRO-TECH). Produced in a wide variety of colored anodized tubes, the AT66 ($359) has essentially identical optics to the William Optics mini-scope. The only significant difference between the two lies with their focusers. The AT66 has a 1.25-in. focuser while the William Optics 'scope has a 2-in. focuser, making it more useful for adding photographic adapters. If you want the widest fields of view with big 2-in. eyepieces the Zenithstar 66 is the better choice. Other than that, choose the model (or color) that's right for you. Alternatively, Sky Watcher and Kunming United Optics also produce competitively priced clones of the ASTRO-TECH 'scope.

William Optics and ASTRO-TECH also market two other ED doublets, slightly larger instruments built around the success of the 66 SD. William Optics produce the Zenithstar 70 (f/6.3) and the Megrez 72 FD (f/6). ASTRO-TECH also market an almost identical 72-mm f/6 ED ($379). Although the images these 'scopes serve up are quite comparable to the Zenithstar 66, the Megrez 72 FD and ASTRO-TECH 72 deliver slightly brighter views. Both 'scopes when kitted out with a 1.25 in. diagonal and eyepiece still weigh in at or just over 4.4 pounds (2 kg), making them easy to use and transport in the field.

Fig. 2.4 The ASTRO-TECH 72 ED doublet (Image © Altair Astro. Used with permission)

As commented on before, the FD labeling on the 72-mm model is a little annoying, especially since it's an ED doublet (most probably FPL-51). A quick daylight look through one of these 'scopes shows that although color correction is very good, it is not as color free as its smaller sibling, the Zenithstar 66 SD. For the record Stellarvue (California, USA) also offers a 70-mm f6 ED 'scope. Called the SV70ED, it comes with all the features of the William Optics Zenithstar 70 but includes a threaded dust cap with the Stellarvue logo, a Vixen-style mini rail and a very nice hard case all for $399.

The Borg Mini-Scopes

When it comes to having fun with a small portable refractor, no company seems to understand the market better than Hutech Corporation and their series of tiny, high quality ED refractors in the range of 1.8–3-in. apertures (45–76-mm). These Mini-Borgs offer a range of finely made 'scopes with Japanese optics. They are modular in design and so can be used with other Borg accessories for visual use, wide field imaging or just for guiding larger 'scopes during long exposure astrophotography.

Perhaps the most remarkable of all is the MiniBorg 45ED, the world's smallest apo refractor. Sporting a high-quality ED doublet objective, this little 'scope has a focal length of 300 mm (f/6.6) and can be used as a faster one (f/5.6) when used in conjunction with a dedicated focal reducer/field flattener. Only 6.6 in. in length with the focuser racked in and weighing less than a pound, this 'scope would get lost in a woman's handbag! Its well-designed helical focuser has very generous back focus (up to 6.5 in.), so it'll work well visually or with a CCD or digital camera.

Fig. 2.5 The world's smallest apo, the Borg 45 ED (Image © Steve Asbury. Used with permission)

Despite its $349 price tag, it's hard not to have a soft spot for this 'scope. Images are crisp and color free, and it's a super little instrument for looking at the Moon at a moment's notice. It'll take magnification well but won't break the laws of physics. Hutech Borg also offers a similarly designed 60-mm f/5.8 ED model for significantly greater resolution and light grasp.

More recently California-based Stellarvue have launched their own mini ED 'scope, which sports a 50-mm doublet objective and a focal length of 330-mm (f/6.6) and can accommodate 1.25-in. accessories for visual or imaging applications. Weighing in at just 2.5 pounds, it comes with a clamshell, dovetail plate and soft carry case. The downside is that you need to spend quite a lot of money to acquire this limited aperture 'scope ($499).

Choice 60-mm Telescopes

The William Optics, ASTRO-TECH and Borg mini-scopes are a good dollar value. They are the Ford KA of small, ultra-portable 'scopes. But some aspire to owning a Mercedes A Class, and in the telescope world there are several candidates: the Takahashi FS-60C and two from Tele Vue (their 60 and 76 models). All are ED doublets of excellent optical and mechanical quality.

Takahashi affectionately calls the FS-60C the 'itinerant' telescope par excellence. This tiny 2.4-in. (60-mm) refractor weighs virtually nothing (OK, 2.9 pounds for the optical tube assembly) and is less than 12 in. long when used in visual mode. Optically, it's a fast f/5.9 doublet with a fluorite front element mated to a low dispersion flint. The little Takahashi excels mechanically as well. Its oversized 2-in. rack-and-pinion focuser is thoughtfully designed for astrophotography and CCD imaging. A thumb screw maintains the focuser in position whatever the direction of pointing is.

You'd expect such a fast doublet to show a bit of color, but some careful testing by day and night shows that views are almost entirely devoid of chromatic aberration. It's one sharp optic. It will handle 150× on a good night before the image begins to go a bit soft. Moreover, by using an adapter called the Extender Q (exclusively designed by Takahashi) the focal length of the native 'scope can be extended from 355 mm to 566 mm (a 1.6× focal length boost), and it'll be easier to achieve high power for lunar and planetary viewing. However, the expensive Extender Q ($268) is probably overkill if you only wish to use it on this tiny 60-mm 'scope. Better to spend your hard-earned money on a high quality eyepiece that'll do the trick. A 2–4-mm Nagler zoom (discussed later) fits the bill perfectly!

Al Nagler, founder of Tele Vue Optics, New York, has enjoyed an almost guru-like status among small refractor lovers, especially in the United States. But it's not just the amateur astronomy community who venerate him. Unlike the other high-end refractor makers, Nagler has vigorously marketed his prestigious mini 'scopes to the birding community. And it's paid off. Tele Vue's two smallest refractors are now as likely to be used by day as they are by night. The smaller of the two, the

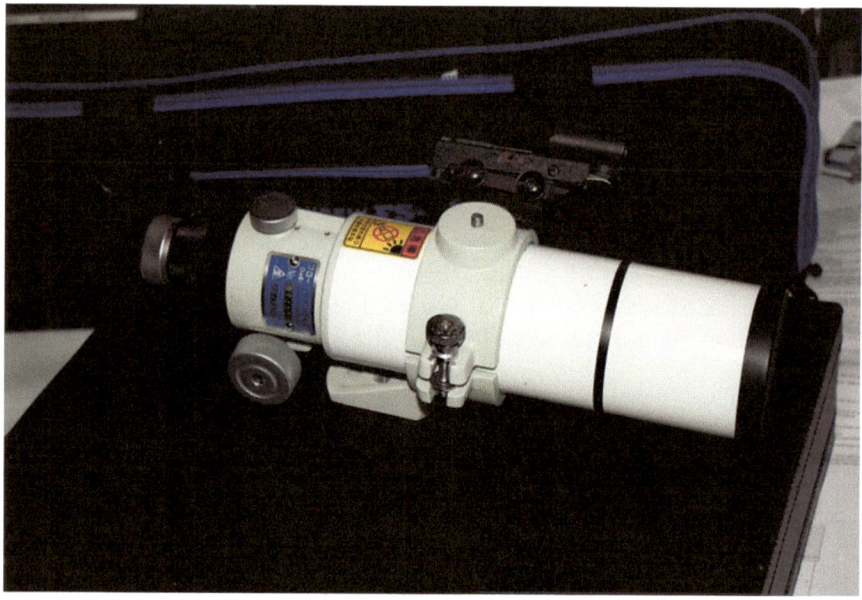

Fig. 2.6 The Takahashi FS-60C (Image © Geoffrey Smith. Used with permission)

Fig. 2.7 The beautiful, tapered tube design of the Tele Vue 60 apo (Image © Venturescope. Used with permission)

Tele Vue 60, is arguably the most beautiful mini-telescope in the world! Optically, it's got a very similar specification to the Takahashi FS-60C, but its mechanical design couldn't be more different. This is a telescope designed for the discerning visual observer who wants to extract the very finest images from an ultra-light portable setup. Its focuser is a 1.25-in. format, so you can't use 2-in. eyepieces with it like you can on the mini-Takahashi, but the Tele Vue 60 can still deliver a maximum true field of 4.3° and a 24-mm Panoptic.

Neither is its focuser a rack and pinion, as you find with the Takashashi FS-60C. Instead, Nagler reverted to the wondrously smooth helical focuser design once used on the now discontinued 70-mm Tele Vue Ranger. Coarse focusing is done by loosening the knob at the top of the tube and sliding the draw tube in and out. When an approximate focus is achieved, the knob is locked and the helical focuser takes over to do the fine tuning. It sounds a bit clumsy—and more than a few folk have expressed dissatisfaction with it—but it's fairly intuitive to use and gets there in the end. After 5 min in the field, you'll have memorized the approximate distance the draw tube needs to be extended for quick results.

Weighing in at just less than four pounds with a diagonal and eyepiece in place and measuring just 10 in. long with its dew cap retracted, it's no wonder Nagler calls it his 'briefcase' 'scope.

More Complex Alternatives

More recently, a number of telescope outlets have introduced 60-mm triplet apochromats. These use not two but three lenses to bring light to a sharp focus. One interesting example, introduced by Altair Astro in the UK is the Lightwave EDT 60-mm f/7 triplet apochromat. Promising excellent color correction—a notch above the aforementioned doublets—this little telescope weighs in at only 2 kg. For £499 you'll get dedicated tube rings, a ten dual speed microfocuser and even a nice aluminum case to pack it all away. The only caveat for long-term use is the relative complexity of the triplet objective, which is more sensitive to misalignment and consequently more difficult to re-center. In addition, some user reports indicated that on high power targets a traditional 60-mm classic refractor delivered more satisfying views, even though the former costs many times less than the latter.

The Altair 60-mm ED triplet has a number of competitors with four elements specifically designed with the astro-imager in mind. Take, for example, the Astrotech AT65EDQ, a 65-mm optical system consisting of four elements in two groups. The first group is made up of a triplet. One of these elements is made from exotic FPL-53 glass with special dispersion properties, which is a key factor for eliminating chromatic aberration. The fourth element is a dedicated field flattener made of ED glass.

As you can imagine, the AT65EDQ ($599), with its four optical elements, is not lightweight. Weighing in at about 2.8 kg (and substantially more when you mate a CCD camera and other accessories onto it), you'll need a fairly beefy mount to get the most out of this for imaging purposes.

These small refractors have an Achilles' heel. All these 60-mm 'scopes, despite being optically perfect, are only 60-mm, and while they handle most daylight projects very well, there are significant advantages to looking for a 'scope with a little more aperture.

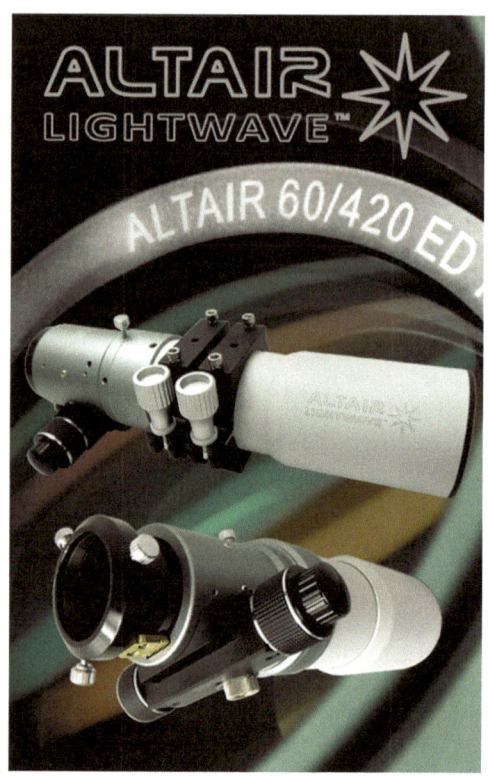

Fig. 2.8 The little Altair EDT 60-mm triplet apochromat (Image © Altair Astro. Used with permission)

Fig. 2.9 The AstroTech AT65EDQ four-element refractor/astrograph (Image © Astronomics. Used with permission)

A Budget Travelscope

No matter which way you slice it, cheap, 'department store' telescopes greatly outsell so-called serious telescopes by a very wide margin. Most people with a casual interest in nature studies or astronomy are not willing to shell out what they see as 'exorbitant' sums of money for a premium telescope and are quite content with a budget model. So when Celestron launched their Travelscope 70 package, this author was keen to see what category this instrument would fall into.

After ordering a presumably random sample from the Internet, the package arrived the very next day. The black carry case containing the telescope and accessories was snugly wrapped inside a double boxed package. The telescope, a 70-mm refractor with a focal length of 400 mm, came with a flyweight photographic tripod, a correct orientation prism diagonal and a couple of eyepieces—20 and 10 mm—delivering magnifications of 20× and 40×, respectively.

Thanks to an exceptionally well written instruction manual, the 'scope was set up in just a few minutes. The total weight of the instrument and mount is less than 2 kg.

The telescope itself is well made and certainly does not create the impression of being toy-like. It has a solid, 1.25-in. metal focuser, which moves smoothly along a simple rack and pinion mode, similar if not identical to that seen on the well regarded short-tube 80 achromat. The telescope is adequately baffled and painted matte black inside to minimize stray light. The finder scope was of very poor quality (possessing a singlet objective) and was not used in subsequent field tests.

After affixing the telescope to the tripod, it was charged with the 20-mm eyepiece, which delivered pleasantly sharp, wide field views. Although it was certainly better than nothing, the tripod proved to be the weakest link. It was just too spindly

Fig. 2.10 The Celestron Travelscope 70 package (Image by the author)

Fig. 2.11 The coated objective of the Celestron Travelscope 70 (Image by the author)

to enjoy prolonged daylight views, where moving the telescope from object to object was a necessity. During a few brief excursions outside at night, this author found maneuvering the flimsy tripod accompanying this telescope to be an exercise in frustration. While it's true that some limited low power sweeps (20×) of the night sky are possible with the supplied tripod, the telescope benefitted greatly by mounting it on something sturdier. In this capacity, the remainder of the field tests were conducted on this telescope my mounting it on a Vixen Porta II alt-azimuth mount and/or a robust table top photo-tripod.

To be honest, great things were not expected from a package costing so little. The objective lens is magnesium fluoride coated. Star testing the instrument using the available eyepieces revealed some astigmatism and field curvature. There was also some obvious spherical aberration apparent as revealed by the asymmetry of the intra- and extra-focal images. Chromatic aberration was obvious using the supplied 40× ocular.

This is not a telescope for high power use—but neither did the instruction manual lay claim to such activities. Making an objective lens well at f/5.7 is not easy! That said, high powers could be coaxed out of the instrument by replacing the stock diagonal with a high quality 1.25-in. dielectric model and adopting higher quality orthoscopic eyepieces. Used with these improved accessories, the magnification could be increased to about 70× before the image began to deteriorate. Stopping down the aperture to 60 mm allowed this author to increase the magnification to 90× or so.

So, just how much fun can you get out of a telescope like this? Quite a bit, as it turned out! The Travelscope 70 makes for a very decent spotting 'scope, so long as magnification is not pushed too high. With a closest focusing distance of 5.8 m (with the supplied 20-mm eyepiece), it doubles up as a nice long distance 'microscope.' In addition, the telescope serves up some decent, low resolution images of

Fig. 2.12 The nicely made rack and pinion focuser of the Celestron Travelscope 70 (Image by the author)

the solar disk using a white light filter; the larger, more prominent spots on the solar photosphere could be resolved with this telescope.

Lunar images through the Travelscope 70 were clean and sharp within the magnification range previously established for it. Our natural satellite presents a wealth of detail that can be studied as its phases change from thin crescent, through first quarter, and on to full Moon. Jupiter's two main belts were cleanly resolved, as well as the large Galilean moons, which change their aspect with the passing of the hours and days.

Very pleasing deep sky views were delivered with an inexpensive 30-mm Plossl eyepiece, delivering a power of 13× and a true field of just under 4°. Sweeping the telescope through the Cassiopeia Milky Way and its rich star fields, one can enjoy nice high contrast views that are only possible with the unobstructed optics of a refractor.

The elliptical shape of the Ring Nebula (M57) in Lyra could be clearly discerned, as well as the famous globular cluster, M13, in the keystone of Hercules, which presented as a small, amorphous blob that was distinctly non stellar at 44×. Using a favorite 8-mm ocular, yielding 50×, this author tracked down a few of the showpiece doubles of the late spring sky, including, Albireo, Mizar and Alcor, Cor Caroli and the lovely orange suns represented by 61 Cygni A and B.

In summary, the Celestron Travelscope 70 was found to be a decent performer within the remit of its very modest price. The light weight of the optical tube would also make it an attractive option as finder 'scope astride a larger instrument. As with any other tool, a telescope is only as versatile as the person using it. And while the instrument would not be recommended as a serious telescope, it is most definitely not a 'toy.' Upgrading some of the accessories will also improve its performance. If you are only casually interested in star gazing and limited to a very tight budget, then this might be a good option for you.

More from Tele Vue

Long-established New York telescope maker, Tele Vue Optics, have offered two very nice doublet refractors for the busy astronomer who demands the highest quality optics in a portable package. These are the Tele Vue 76 ($1,995) and Tele Vue 85 ($2,425).

Introduced in 2002, the Tele Vue (TV) 76 was the replacement for their older ED 'scopes—the Ranger and Pronto—both of which were splendid 70-mm f/7 doublets with good but not apo-quality color correction. The TV 76 (f/6.3) has a slightly larger aperture but the same focal length as the older 'scopes. Like the Tele Vue Pronto, it has a beautiful rack and pinion focuser in a 2-in. format. Bought new, the package includes a custom made soft case, a screw-on lens cover and sliding dew shield, a 20-mm Tele Vue Plossl eyepiece and 2-in. Everbrite diagonal, a 1¼″ adapter (all with clamp ring fittings) and a manual signed by Uncle Al himself. When outfitted with an eyepiece and a diagonal, it tips the scale at just over 6 pounds. That's significantly heavier than some top-of-the-range spotting 'scopes but not enough really to present problems in the field. Any loss of portability, though, is made up for by the TV 76's amazing versatility. A 3-in. aperture is just about large enough to make high resolution visual observing worthwhile, and its short focal length (480 mm), coupled to a big, wide-angled eyepiece, means that you get majestic 5.5° views of the night sky.

The optics on these telescopes are first rate. In comparison to several lower cost ED doublets of similar specification, it is clear that the TV 76 bests them all.

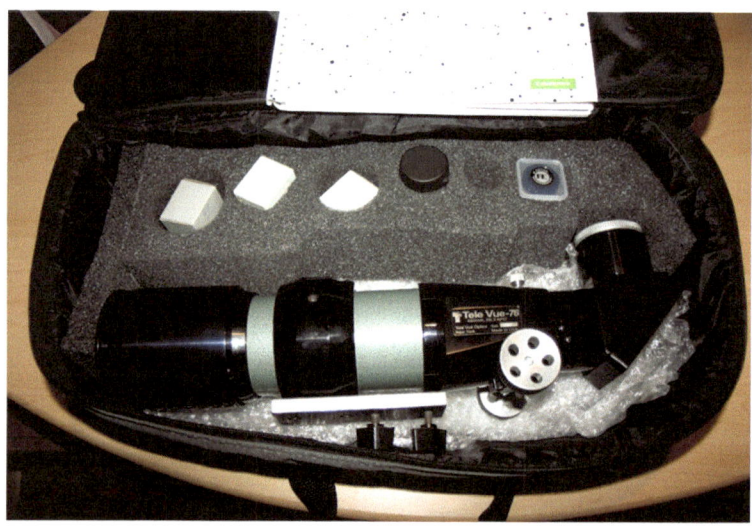

Fig. 2.13 The TV 76 goes anywhere at a moment's notice (Image by the author)

Star testing this 'scope at 120× shows how superbly crafted the optics are. Vega displays a hard white Airy disc surrounded by a single diffraction ring. No significant color error was noted. The diffraction patterns both inside and outside focus were nearer to perfection than in any other telescope. They're cleaner and easier to see compared to the slightly fuzzier patterns usually observed with cheaper ED doublets. Like other two-element apos, it does display a small amount of color on the rim of the diffraction pattern both inside (magenta) and outside (green) focus, but that's normal behavior for an instrument with an ED doublet objective.

It's easy to test good optics, and you don't need an optical test bench to do it. A well-figured lens ought to be able to take very high magnifications before noticeable image breakdown occurs. Daylight and nighttime tests with high-quality eyepieces and image amplifiers show that the TV 76 can take amazingly high powers, and this little 'scope can hold 100× per inch of aperture under ideal conditions. It has very low spherical aberration and is devoid of astigmatism and coma. This is extraordinary for an ED doublet with such a fast focal ratio (f/6.3). And it's no accident either. It's down to the excellent figure of the lens and the employment of a large air gap between the objective elements.

Fig. 2.14 The TV 76 atop a lightweight Tele Vue panoramic mount (Image by the author)

Fig. 2.15 The butter-smooth rack and pinion focuser on the Tele Vue 76 (Image by the author)

The TV 76 really rocks when it comes to resolving the prettier double stars, despite its fairly short focal length. However, like all other short focal ratio 'scopes, maintaining sharp focus can be a bit fiddly, especially during high-power applications under less than perfect seeing conditions. The instrument's excellent color correction makes seeking out variegated doubles a joyous adventure. Albireo, 61 Cygni and Almaak unveil their austere beauty at moderate and high powers. Forget Polaris and Rigel too. These high-contrast companions usually cited as tests for a 3-in. refractor are too easy for this refractor.

More challenging (and more fun) is the lovely triple system of Iota Cassiopeiae, and close binaries such as Delta Cygni and, both of which the TV 76 managed to resolve. And though it's not the hardest binary system to discern with a good 3-in. refractor, Epsilon Bootes (Izar) is arguably one of the most compelling sights to see in a small telescope in all the heavens. Steady skies and high magnifications are required to elucidate its lovely secret—a magnitude +4.6 blue–green companion separated from its primary by just 2.9 arc seconds of sky. Now this little telescope can resolve pairs as close as 1.5 arc seconds *provided* they are of fairly equal brightness. But the near sevenfold difference in brilliance between Izar and its main sequence companion renders the secondary hard to see, overwhelmed as it is by the light of its primary. Optics plays a role with this system, too. The 4-in. instruments consistently struggle with this system, but a high-quality 60-mm refractor should just do the job under good conditions. And one of the finest views of Izar (Epsilon Bootis) can be seen with this 3-in. refractor. During a vacation to a tiny coastal resort on the North West coast of Scotland, this author chanced upon some fair weather. Tucked away in a shallow inlet, the early evening winds subsided gradually to a dead calm after midnight, allowing me to take advantage of exceptional observing conditions with dark magnitude +6.5 skies. On two successive nights, I was able to rack up the power on my 'scope to 200× to get a razor-sharp separation of the system. Like a budding yeast cell seen under a microscope, the pale blue ball of the secondary sat on an otherwise perfect first diffraction ring of a golden orange primary. It's at moments like this that one can more fully appreciate why the famous double star observer Otto Struve christened it *Pulcherimma*!

Its bigger brother, the Tele Vue 85 is everything the 76 is, only larger. It has acquired a rather cult-like following among refractor aficionados. But with its significantly higher price tag, it ought to!

Budget 80-mm Doublets

If you are in the market for a small telescope that could be used at the drop of a hat, then you should consider an 80-mm refractor. This aperture is about the minimum that can engage a serious observer on most objects. Thankfully, there are many varieties available, and here we shall discuss some of the more popular models used by amateur astronomers.

Some of the best bargains on the market today are the 80-mm f/11 achromats, made by Vixen, as well as a huge number of older used models available. Their long focal length, good color correction and quick cool down time have endeared them to lots of people. Excellent for high resolution work like double stars, the planets and the Moon, they can take surprisingly high magnifications and are an excellent alternative to the shorter focal length 80-mm refractors with low dispersion glass.

Fig. 2.16 An 80-mm refractor, an f/11 achromat with a nicely made objective from Nihon Seiko (Image by the author)

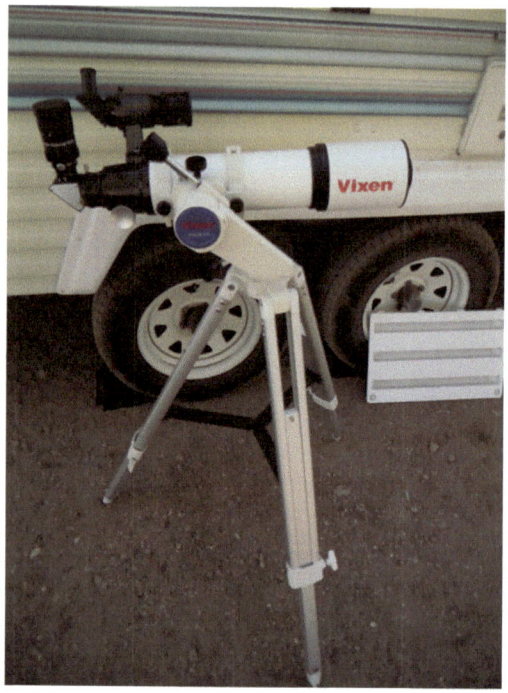

Fig. 2.17 A Vixen 80-mm f/7.5 ED refractor fitted with an 8–24-mm zoom eyepiece (Image by the author)

Fig. 2.18 The SkyWatcher Equinox 80 ED doublet (Image © Sky Watcher. Used with permission)

Perhaps the most popular grab 'n' go refractor of all is the ED 80. Introduced more or less simultaneously by SkyWatcher, Celestron and Vixen, these telescopes sold like hotcakes when they first hit the market in 2004. Sporting an ED doublet objective for improved color correction and a focal length of 600 mm (f/7.5), it is relatively short and lightweight, lending itself easy to transport and use on a fairly lightweight mount. Optically, these telescopes score high, with their sharp, color free images and their ability to be pushed to high powers when called for. Their modest price ($599) makes them a very attractive package for the savvy grab 'n' go astronomer.

Since then, a number of similarly designed 80-mm ED 'scopes have appeared under Chinese rebranded names such as Altair, Barska, Ikharus and Sharpstar.

The 80-mm Triplet Apos

As well as offering their smaller refractors in the 66- and 72-mm aperture classes, William Optics also market two slightly larger but still very portable ED triplet refractors. The first offering is the GTF 81 ($1,219 for the tube assembly only). Sporting a focal length of 478 mm (f/5.9), it comes with a sturdy, 2.5-in., fully rotatable focuser. The larger aperture model—the FLT 98—offers slightly better performance, owing to its larger aperture, but it comes at a very steep price ($2,495)—fully twice that of the smaller GTF81.

Other companies offer very similar products. Take, for example, the Meade Series 6000 triplet apochromats. The 80-mm utilizes the high-grade fluorite-based FPL53 glass in its lens construction, which is not present in its larger models. In a very attractive package deal, the telescope comes with a 2″ Series 5000 Enhanced 99 % Reflectivity Slip-fit diagonal, aluminum cradle rings with mounting dovetail and hard case. All this for $999.

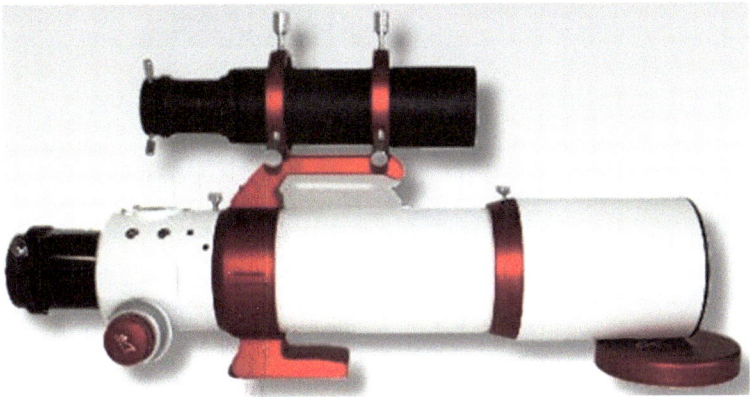

Fig. 2.19 The William Optics GTF81 triplet apo (Image © OPT. Used with permission)

Fig. 2.20 The 80-mm Meade Series 6000 triplet apochromat (Image © OPT. Used with permission)

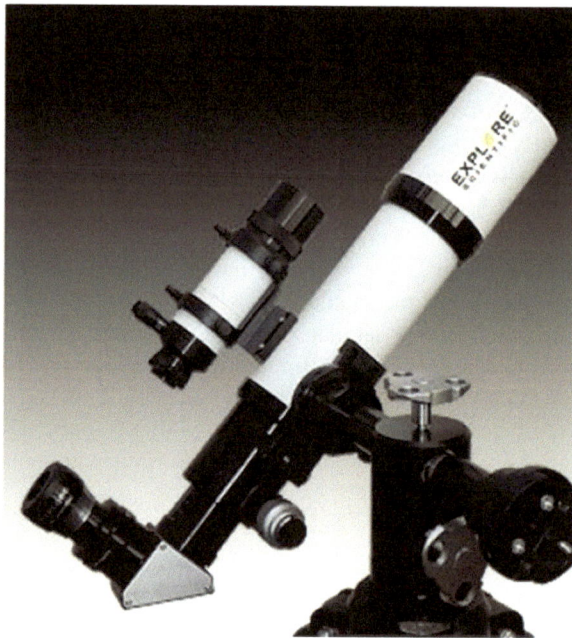

Fig. 2.21 The Explore Scientific 80-mm triplet apo (Image © OPT. Used with permission)

If you think Meade is offering a good deal on its 80-mm Series 6000 triplet refractor, take a closer look at the Explore Scientific equivalent, which offers much the same package for $799. Unlike the Meade model, the Explore Scientific 80 mm does not come with a carry case but offers the same optics in a lighter weight carbon fiber tube for an extra $200.

With optical tube weights averaging about 7.5 pounds, these telescopes are very easy to transport in the field. Their triplet optics ought to deliver excellent images but are more sensitive to misalignment and temperature fluctuations than their doublet counterparts.

Older Models to Look Out For

Although the market is flooded with new models promising the Earth, it is reassuring to know that some excellent 3-in. grab 'n' go telescopes can still be had on the used market. Arguably one of the best is the Takahashi FS 76 doublet apochromat. Now discontinued, this small refractor offers up razor-sharp, sharp and color-free images a notch up from the more ubiquitous ED 80 refractors.

Fig. 2.22 The superlative Takahashi FC76 (Image by the author)

In the achromatic genre, keep a look out for a used Stellarvue 80/9D. Build quality is excellent in an all-metal tube, 2-in. focuser (rack and pinion) and a retractable dew shield with a screw on metal dew cap. Star testing on Vega showed that the optics were undiminished, that is to say, excellent. An 80-mm f/11 achromat, mounted side by side with the SV80/9D, provided a good test. After checking on a few stars; the best available of which included Altair and Deneb, it was pretty clear the latter had a distinct edge optically; the star test marginally better and in-focus star fields displaying slightly greater contrast. Two fine 'scopes, one just a wee bit better than the other! The SV80/9D is one of the sweetest 80-mm achromats available. It was well worth any additional cost of cleaning it up. Kudos again to Vic Maris of Stellarvue. Indeed, as you'll later discover, this modest telescope was used to divine many of the showpiece objects in the sky, described later.

Fig. 2.23 The Tele Vue Gibraltar mount is beautifully matched to the SV80/9D (Image by the author)

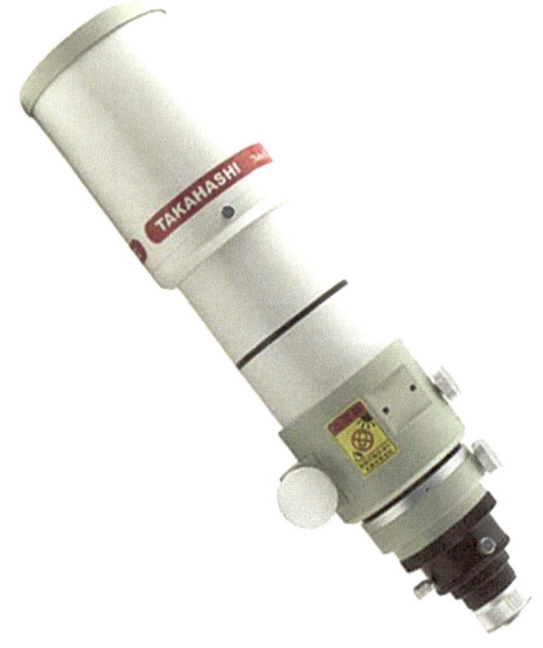

Fig. 2.24 The Takahashi FSQ-85ED "Baby Q" (Image © Astronomics. Used by permission)

Takahashi, in Japan, has also introduced its own version of the ultra-portable imaging 'scope in the corpus of the FSQ-85ED. Dubbed the "Baby Q," this 85-mm four-element apochromatic refractor sports top notch optics but at a price that will take your breath away ($3,650). Are these four-element designs for you? Perhaps, especially if you have imaging in mind. But for purely visual pursuits, simpler designs are more reliable and will serve up equivalent or better images at a significantly lower cost.

A Very Special 90-mm Achromat

Excellent grab 'n' go refractors needn't cost a fortune. Take the SkyWatcher Evostar 90-mm f/10.1 achromat, which can be bought used for much less than the original cost.

The telescope has a lovely, lightweight tube, and is thus eminently portable. The dew shield is plastic, which is not a problem in itself. The dust cap (also plastic) is neat looking with a 60-mm aperture mask, enabling it to operate as a 60-mm f/15. Cool!

Fig. 2.25 The SkyWatcher Evostar 90-mm f/10.1 achromat (Image by the author)

Fig. 2.26 All good grab 'n' go 'scopes need a lens cap (Image by the author)

Removing the dew shield reveals a pristine 90-mm air spaced doublet objective with very nice multi-coatings applied to all surfaces. The instrument cannot be collimated, however. Checking collimation with a Cheshire eyepiece under a bright blue sky revealed good but not perfectly aligned optics, but neither is it enough to worry about in the slightest. Unlike some earlier budget achromats examined, the paintwork on the inside of the tube was immaculately applied and the baffling is more than adequate.

The focuser is the regular 1.25-in. Synta rack and pinion. Nothing to write home about, but more than adequate for the task; remember, it's f/10.1! It handles heavy 1.25-in. eyepieces very well. Pity it doesn't come with a 2-in. focuser for use with modern wide angle eyepieces.

This 'scope can be used with a regular 90° diagonal for astronomical use or a 60°, correct orientation diagonal for terrestrial applications, like this one.

Daytime assessment of the optics revealed nothing out of the ordinary. Indeed low power views of the birds seeking refuge in the trees nearby were sumptuous—tack sharp and rich in detail, free of any secondary spectrum.

Inserting a medium-power eyepiece (and filter, of course), you will be delighted with the sharp, high resolution detail you can see on the solar photosphere.

Fig. 2.27 The fully multi-coated objective of the SkyWatcher Evostar 90 (Image by the author)

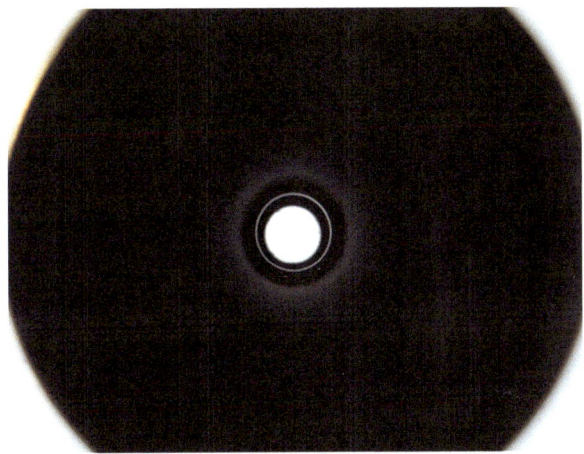

Fig. 2.28 The SkyWatcher Evostar 90 is nicely baffled and blackened internally (Image by the author)

Contrary to what you may have heard before, it only takes one good night to assess a telescope's optics. Only on imperfect nights does one need to pass it through multiple star tests. Centering the bright star Vega in the field of view of an eyepiece generating 150×, the Evostar produced a clean, white, sharply focused

Fig. 2.29 The simple rack and pinion focuser on the SkyWatcher Evostar 90 (Image by the author)

Fig. 2.30 The SkyWatcher Evostar can be fitted with a 60° diagonal for terrestrial viewing (Image by the author)

Airy disk with a nicely formed first diffraction ring. The star is surrounded by a beautiful halo of purple; absolutely normal for this type of telescope and perfectly acceptable for visual use. Defocusing ever so slightly revealed no signs of astigmatism. Examining the Fraunhofer diffraction pattern both intra- and extra-focally using a green filter demonstrated smooth, well corrected optics with only slight under correction. All in all, a very good result for a telescope that costs so little!

Epsilon 1 and 2 Lyrae can be perfectly resolved into four stars. The greenish companion to Epsilon Bootis—now sinking lower into the western sky—was nicely separated from its orange primary at 150×. The same was true of Delta Cygni. Pi Aquilae, while certainly not resolved, was plainly shown to be strongly elongated. Its deep sky prowess was decent, too. During the darkest period of twilight, this author could see the beautiful incandescent annulus that is the Ring Nebula, M57, sharply defined at 101×. Moving into Hercules, one can track down the globular cluster M13 and be rewarded with a bright, highly condensed image. With a concentrated gaze, the attentive observer can make out quite a few individual stars on its periphery. The telescope does not disappoint on open clusters like M52 either, its modest aperture revealing several dozen stars in a kidney-shaped arrangement, the eye being drawn to a striking eighth magnitude orange sun on its flank. The lack of field curvature of this 'slow' object glass makes examining these bright clusters a particular joy.

In summary, the SkyWatcher Evostar 90-mm achromat is arguably one of the best buys in the small refractor market. Usually touted as a beginner's telescope, it is much more than that. The Evostar 90-mm would embarrass instruments costing ten times its modest cost new. It is super light, super portable and delivers fine views at low and high power. If you're in the market for a no-nonsense 90-mm telescope, this would be the one to go for. Because of its low cost and excellent durability, it would be especially useful to astronomy clubs undergoing public outreach. It would also serve as an excellent telescope for the instruction of undergraduates in astronomy/astrophysics.

The 4-In. Refractors

What is the most versatile refractor for grab 'n' go purposes? Well, such an instrument ought to be large enough to gather a decent amount of light to enable the observer to engage with both the deep sky as well as Solar System objects, including high resolution images of Luna and the bright planets. There is arguably one instrument that can fit that niche supremely well; enter the 4-in. refractor.

Like the smaller refractors, there is a huge range to select from, both achromatic and apochromatic. The biggest recent buzz has come from the Celestron 102 GT, which was being offered by several telescope outlets for significantly less than $100.

These are 4-in. (102-mm) achromatic doublets with a focal length of 1,000 mm. At f/10, the color correction is quite acceptable, allowing you to push magnifications to 200× and more. These Chinese-made instruments are perhaps the best testimony yet to the great virtues of traditional achromatic optics. They will deliver razor-sharp views of the Moon and planets, double stars and the brighter deep sky objects. A great many people bought them out of sheer curiosity. How can you get something so good for so little? The answer lies in the relative ease with which modern production methods can churn these objectives out both cheaply and well.

Fig. 2.31 The Celestron 102 GT (Image © Celestron. Used with permission)

Of course, you still have to outfit the Celestron 102 GT with a star diagonal and tube rings and mount it decently, but it still ranks as one of the best deals this author has witnessed in decades.

If you're in the market for the classic looks of a long refractor that is still very portable and quick to cool down, then look no further than the Astrotelescopes 102-mm f/11 achromat. Manufactured by the Kunming Optical Company in China, the author contacted the sole UK importer of these instruments, Lyra Optical, Rochester, and requested a sample for review. Supplied as an optical tube assembly, the only accessory included in the package was a pair of well-made tube rings.

Tipping the scales at just over 10 pounds, the nicely finished white CNC tube is fitted with a fully retractable dew shield. Fully extended, the tube measures 47 in. long but, impressively, is 8 in. shorter when collapsed away for storage. That makes it shorter than a 4-in. f/10 achromat!

After pointing the telescope skywards, it was discovered that the dew shield was held too loosely, and it fell back down the tube with a loud 'thwack.' The problem can be remedied by affixing some blu-tack to the sleeve. This could have been easily avoided by padding it with more felt.

The telescope comes with a solidly made, fully rotatable, two-speed Crayford focuser that can accept 1.25- as well as 2-in. accessories. In practice, the focuser seems to be a little fiddly. Specifically, it had a bit too much tension and made fine focusing slightly more difficult to achieve in comparison to other units.

Fig. 2.32 Despite its longer focal length, the 4-in. f/11 (*lower*) collapses to a shorter tube than its f/10 counterpart (*upper*). The tape measure is set at 110 cm for scale (Image by the author)

The objective is a fully multi-coated, air-spaced crown-flint achromat. The coatings look smooth and very evenly applied. The inside of the tube is painted matt black, and three baffles keep stray light at bay. A telescope this size is easily mounted equatorially or on a simpler alt-azimuth mount. Both systems were used to evaluate the performance of the telescope in the field.

This telescope generated high expectations in terms of optics. With its high F-ratio, it ought to have a number of desirable features for the visual observer. For example, it ought to enjoy greater areas of the field of view that are diffraction limited. Its deep light cone should also allow inexpensive eyepieces to work better, especially in regard to their ability to reduce astigmatism. The gently curving f/11 optics ought to be easier to make in comparison to its shorter F-ratio counterparts. What is more, high F-ratio 'scopes are easier to collimate and less sensitive to misalignment. In addition, the greater depth of focus of high F-ratio telescopes renders them much easier to focus accurately. That's why, by necessity, short F-ratio (fast)

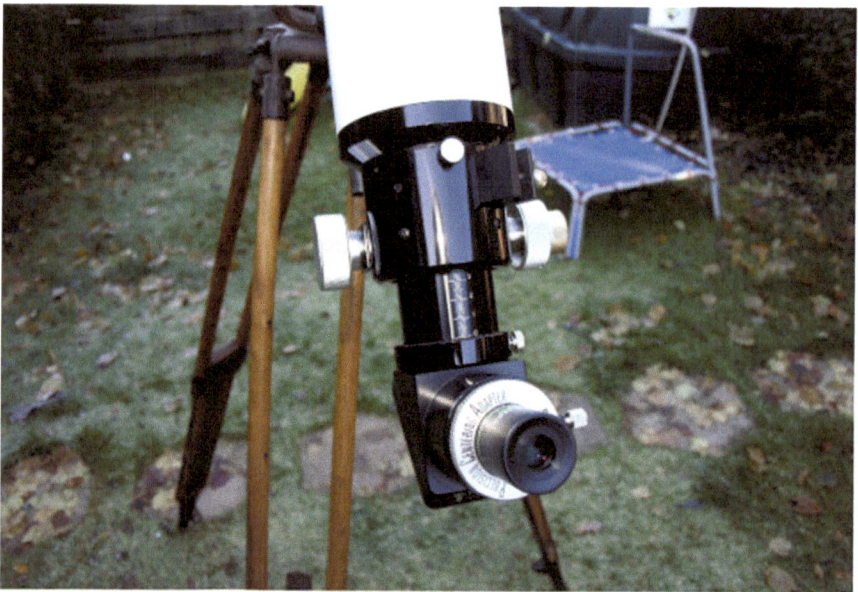

Fig. 2.33 The two-speed Crayford focuser is basic but adequate. It would have been preferable to have a simpler rack and pinion with this instrument, though (Image by the author)

Fig. 2.34 The multi-coated objective of the 4-in. f/11. Note the three light baffles inside the tube (Image by the author)

'scopes require micro-focusers. Furthermore, the ordinary crown and flint glass changes its shape less severely than a typical ED 'scope of the same aperture (an important attribute as temperatures continue to fall during an observing run), allowing you to get diffraction-limited images faster than the latter.

So how did this 4-in. f/11 live up to these considerations? As it happens, very well indeed! Mounted on both a Gibraltar Alt-Az as well as a sturdy equatorial mount, the 'scope was subject to field tests over several weeks of extremely cold weather. Star testing at 167× showed very well corrected optics. In focus, bright stars such as Vega and Capella focused to nice round Airy disks. There was no sign of coma or astigmatism. The stars were surrounded by an unfocused violet halo, a natural consequence of the telescope's achromatic nature. Racking the 'scope inside focus revealed textbook perfect diffraction rings devoid of rough zones. In that scenario, the focuser was ever so slightly cutting off a bit of the light cone and impacting the edges of the unfocused stellar image.

Extra-focally, the rings were just a tad softer, a classic sign of slight under correction. Inserting a 40-mm wide angle Erfle delivering a wide 2.4° field of view, a small amount of distortion in the outermost 10 % of the field could be detected, while the same eyepiece showed up a perfectly corrected field in a 4-in. f/15 instrument set up alongside it. All in all, this optic rates very high, certainly a notch up on the very capable 4-in. Russian f/10 achromat but not quite in the same league as the 'sensibly perfect' f/15 achromat.

With the testing over, it was time to relax and enjoy using the 'scope. The first target was a post opposition Jupiter. During good seeing conditions the 4-in. f/11 delivered very sharp, high contrast images of the giant planet, with several bands evident and hints of ovals and swirls set among them. The slight violet halo was actually quite pleasing to the eye. Certainly, it did not detract from the fine views this 'scope delivered. The first quarter Moon was a sight to behold at all magnifications from 28× to 200×. The 'scope's mettle in subzero temperatures was tested on a suite of difficult multiple star systems, including Delta and Mu Cygni (triple system), Theta Aurigae and Eta Orionis. It passed them all with flying colors. The generous 2.4° field offered with a 2-in. eyepiece presented the glories of M35, the Double Cluster (C14), the Alpha Persei Association and the Orion Nebula (M42), in rich, high-contrast detail.

Minor mechanical issues aside, this telescope appears to be a viable alternative to a 4-in. ED 'scope (and at less than half the price). It shows more color than an ED 'scope, no question, but an experienced observer should be able to divine everything the ED 'scope can. So, it's not just a very good 'scope for the money, it's simply a very good 'scope. While other examples of this 'scope may or may not prove to be as high quality, this Chinese 4-in. f/11 achromat was certainly true to the promises of its design. Isn't it nice when optics and experience sing from the same hymn sheet!

A Russian Favorite: The Tal 100RS

Introduced in the late 1990s by the Novosibirsk Instrument-making plant in Russia, the Tal 100R enjoyed a lot of success, especially in Europe, owing to its razor-sharp images with minimal false color and little in the way of other aberrations that can

Fig. 2.35 The Tal 100RS refractor astride a lightweight Vixen Porta mount (Image by the author)

ruin an image. Over the last decade, they've had consistently good optics. But the original Tal 100R had its problems. For one thing, the mount accompanying the instrument, despite its good-looking appearance, was insufficiently robust to do justice to a fairly long, 4-in. refractor. The telescope itself had a 1.25-in. focuser with very limited focus travel. Some eyepieces couldn't come to focus with it; and as for using a Barlow lens—forget it! In addition, while the supplied 1.25-in. diagonal is of high quality, its peculiar design means that it cannot be interchanged with other 1.25-in. diagonals from other manufacturers. Finally, a number of users had issues with the inadequate internal baffling of these 'scopes. But once these issues were sorted out, the instrument performed excellently.

This author compared the Tal 100R achromat with a Tele Vue f/5 Genesis fluorite refractor and found that the Russian achromat had the edge on planets and double stars. Over the years since its introduction, the Tal 100R has undergone some significant changes. For one thing, the 1.25-in. focuser was replaced with a solid 2-in. focuser (which looked very 'Synta esque'). But it had a great deal more focus travel and so could now be used for almost any visual or photographic application. The reassuringly solid aluminum tube finished in white was totally redesigned with a vastly superior baffling system. The new model was renamed the Tal 100RS.

There are many other 102-mm (4-in.) refractors to choose from on the market. The only real reason to acquire these instruments is for their improved color correction, but the differences between a good f/10 achromat and the former are subtle at best. That said, some very nice instruments in this aperture class need mentioning. First up is the Tele Vue102. This venerable 4-in. refractor is considered by many to be the ideal grab 'n' go instrument. It's large enough to collect enough light to keep a serious observer busy for years, and can serve up views of the Moon, planets and double stars rivaling those seen through much larger reflectors.

This author's first serious foray into the world of high-end 4-inchers came when he acquired a 'classical' Tele Vue Genesis refractor (non-SDF). It was a beautiful 'scope, built like a tank and capable of producing wondrous, sharp images with a very flat field. Low power views more than 5° wide were awe-inspiring. It was also a solid performer on Luna and bright planets, taking magnification well and only showing a little bit of false color on high-contrast objects when pushed above 150× or so. It was close to a 'perfect' telescope, but a few things niggled me about it. For one thing, because it was such a fast refractor, finding the sweet spot of fine focus was often a challenge, especially when used at high powers. When you have an f/5 refractor, there's no room for ambiguity; you're either in focus or you're not. In addition, it had a very short focal length—504 mm—and so was difficult to get ultra high power views. Thus, added to a four-element objective, one has to resort to adding a two-element Barlow lens and a five-element Plossl eyepiece to get those savored 200×+ views that are achievable on the steadiest nights. That's a lot of glass to put between you and the heavenly realm, so much so that I wondered whether it would really take the edge off many high resolution images of the Moon and planets. After briefly looking at a variety of double stars, Mars and Saturn through the economical Orion 100ED (a f/9 FPL-53 ED doublet), it was immediately convincing that the Genesis was indeed failing to deliver the best high power views.

The Tele Vue 102 (now discontinued) is a doublet ED refractor with a aperture of 102 mm (4 in.) and focal length of 880 mm (f/8.6). As you might expect, the instrument comes with a excellent rack and pinion focuser and retractable dew shield. The tube is beautifully made of high strength aluminum, finished in a textured ivory powder coat. Optically, it leaves little to be desired, throwing up excellent high contrast images of the Moon, planets and brighter deep sky objects. Star testing a unit revealed excellent spherical correction with no signs of astigmatism or coma that plagued a refractor image. The 102 is an excellent double star telescope being capable of resolving pairs down to its theoretical limit.

More recently, Synta introduced the ED 100, a very similar performing instrument to the Tele Vue 102 but at a significantly lower cost. Optically, there is not much to choose between them but the heirloom quality of the Tele Vue mechanicals

Fig. 2.36 The lovely Tele Vue102 atop a wooden Gribraltar Alt-Az mount (Image by the author)

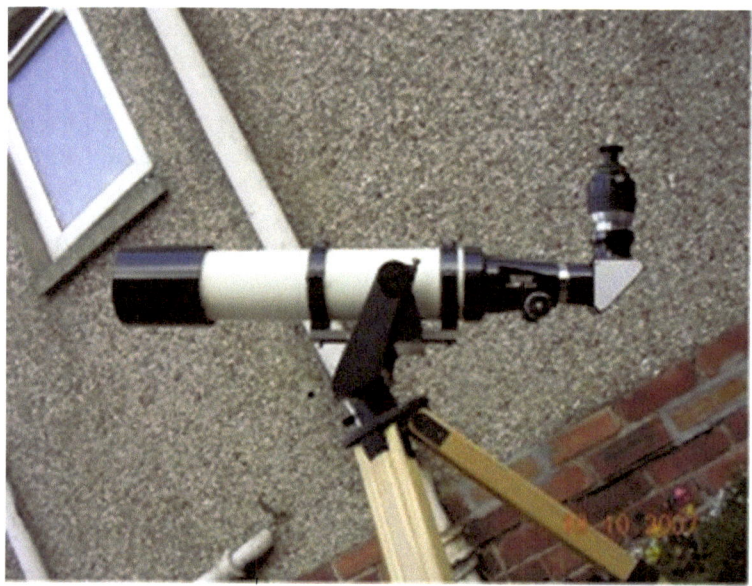

Fig. 2.37 The original Tele Vue f/5 Genesis on a Gibraltar mount (Image by the author)

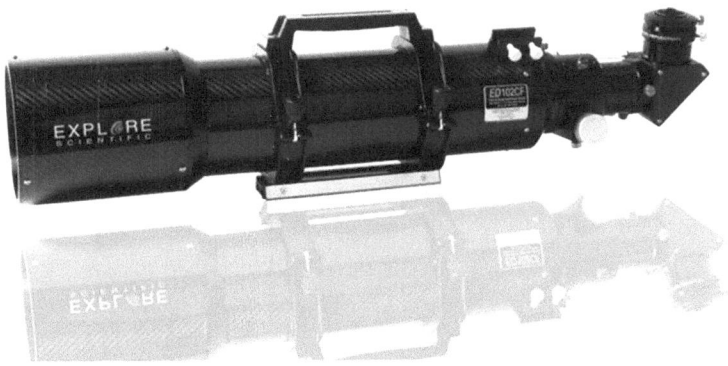

Fig. 2.38 The Explore Scientific carbon fiber 102-mm triplet apochromat (Image © OPT. Used with permission)

leave the basic ED 100 model in the dark. That said, many ED 100 users have elected to upgrade the basic focusers that accompany these instruments, and an army of loyal fans use them each and every clear night.

Those on the lookout for the most compact, ultra Portable 4-inchers would do well to look at the current suite of triplets and four-element Petzvals. Because of their greater complexity, they invariably command a greater price than the simpler models outlined above. Also, in the long run, their optical components will have a tendency to go out of mutual alignment, hence requiring readjustment.

Chinese-made models, such as the Explore Scientific 102 mm f6.7 ED triplet refractor ($1,349), feature a nicely designed objective using Hoya low dispersion glass, a state of the art dual-speed 2-in. focuser, tube rings, mounting plate and a built-in carry handle. Other variations on this theme are marketed by Meade Instruments.

Like the smaller Explore Scientific triplets, these telescopes come with everything you need to get started—a 2-in. dielectric diagonal, tube rings, a dovetail plate and an ergonomic hard case.

Triplet apochromats, like any other type of instrument, come in various tiers of quality. Here, we'll showcase just one 'high-end' model in this aperture class, the APM 107/700 F6.5 Super ED.

With a lens designed by Russian optics giant LZOS, the objective has a fully multi-coated FPL53 ED element ensuring excellent color correction. This model comes with a very robust 3-in. fully rotatable dual speed focuser to allow heavy accessories like large eyepieces and CCD cameras to be attached. The 3-in. focuser

Fig. 2.39 The APM 107/700 F6.5 Super ED triplet apo (Image © Altair Astro. Used with permission)

bumps up the mass of this unit to 5.7 kg. With a retractable dew shield, the instrument collapses to a very Comfortable 580 mm and when fully extended 655 mm. A hard carry case is included in the price tag.

The Last Word: Why Not Consider a Rich Field Refractor?

Thus far, we have only considered achromats and apochromats that offer excellent color correction. But there are also a number of large aperture rich field telescopes available at very reasonable prices. These are usually simple achromatic doublets with short focal lengths, optimized for deep sky viewing at low and moderate powers. Because they display fairly prominent chromatic aberration on bright objects, they are not the best instruments for lunar and planetary study but can be used for casual observing. Here, we will discuss a couple of models that have piqued the attention of the amateur community.

First up is the Explore Scientific AR 102 package. The telescope has a 4-in. (102 mm) aperture and is fully multi-coated, possessing a focal length of 650 mm. The AR 102 has a robust 2-in. dual speed focuser to accommodate the sturdiest of (often) heavy wide angle oculars, a 8×50 finder tube and rings with carry handle dovetail plate. The package also included a high quality 2-in. dielectric diagonal.

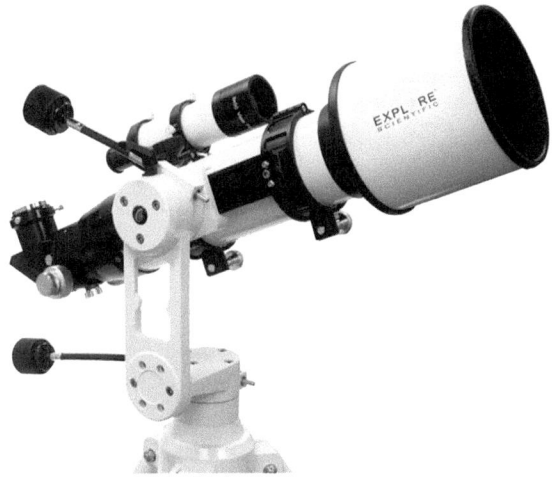

Fig. 2.40 The Explore Scientific AR 102 (Image credit and © OPT)

A similar 4-in. f/6.5 doublet achromat was put through its paces. Its performance was surprisingly good on all celestial targets. Magnifications could be pushed to about 160× for lunar and planetary observations and indeed can be extended to nearer 200× with an appropriate minus violet filter. This author has enjoyed many hours with one of these instruments and after experimenting with filters that suppress chromatic aberration, found that they were decent instruments on tough targets like Jupiter, Mars and the Moon. Where a telescope like this really shines is in the deep sky work, where its relatively large aperture pulls in some of the fainter galaxies, globular clusters, open clusters and bright emission nebulae.

If you want a telescope complete with a computerized go-to mount then you might want to give the Celestron Nexstar 102 SLT a closer look. Featuring optics with very similar specifications to the AR102, the telescope comes with a light, but fairly sturdy, computerized mount. With the touch of a few buttons, you can slew it off to any of 4,000 objects, where the telescope will center the object of interest and allow you to automatically track it for half an hour or more. Two eyepieces delivering 26× and 73× allow you to make a decent start, but the experienced user will almost certainly upgrade these budget oculars to higher quality examples to get the very most out of the very decent optics delivered by this telescope. All in, such a complete package will set you back a mere $430 plus shipping.

This brings to an end our brief survey of the grab 'n' go refractor market. As you've seen, there is an enormous range of instruments to suit everyone's needs and budgets. Choose the model that best suits your schedule. Of course, if you wish to

Fig. 2.41 The Celestron Nexstar 102 SLT (Image © OPT. Used with permission)

learn more about refractors specifically, then check out *Choosing & Using a Refracting Telescope* by this same author for more details. In the next chapter, we'll take a look at the world of the Newtonian reflector, a long time favorite of the grab 'n' go astronomer.

Chapter 3

Grab 'n' Go Newtonians

The basic design of the Newtonian reflector—so named because of its invention by Sir Isaac Newton—has hardly changed much since it was conceived of by the great scientist back in the year 1668. Instead of using a system of lenses to focus light, Newton hit on the idea of using a finely polished spherical mirror to collect and focus the rays of light from distant objects.

These days, the primary mirror is not spherical but parabolic. That slight change in shape means that the all of the rays parallel to its axis will be focused at the same point. The substrate out of which the primary mirror is fabricated tends to be Pyrex or plate glass, and since it only needs to be figured and aluminized on one surface, it makes their manufacture significantly easier than the refractors described in the previous chapters.

Not surprisingly, there exist a nice variety of commercial Newtonians in apertures ranging from 3 in. (76 mm) up to 6 in. (152 mm) that can rightly be considered grab 'n' go 'scopes. The remainder of this chapter is devoted to describing some of the more popular models.

In 2009, to celebrate the International Year of Astronomy, a number of very inexpensive but functional reflecting telescopes were brought to market. Tiniest of all is a 3-in. instrument called the Celestron Firstscope package ($44.95). This demure Newtonian has a focal length of 300 mm (f/3.9). So what do you get for this? Well, the telescope comes with two eyepieces delivering powers of 15× and 75×, powerful enough to get decent views of the Moon, wide pretty double stars and the brighter deep sky objects.

N. English, *Grab 'n' Go Astronomy*, The Patrick Moore Practical Astronomy Series, 53
DOI 10.1007/978-1-4939-0826-4_3, © Springer Science+Business Media New York 2014

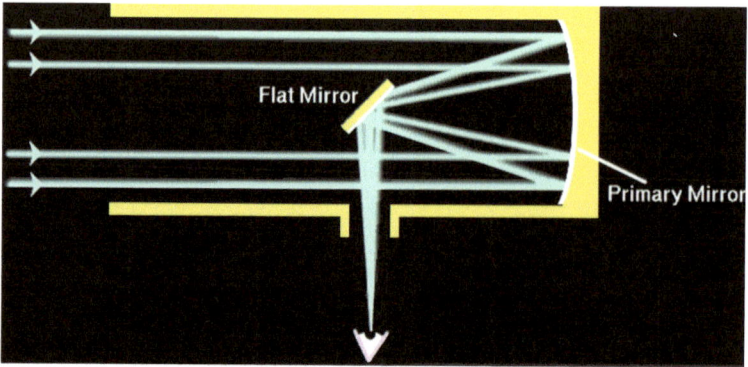

Fig. 3.1 Diagram showing the basic principles of a Newtonian reflecting telescope (Image © Oldham Optical. Used with permission)

Fig. 3.2 The Celestron Firstscope (Image © OPT. Used with permission)

As you can imagine, the Firstscope is ultra portable and great as far as storage is concerned. It looks cute anywhere, but can it deliver? The answer is most definitely yes, so far as you don't set your sights too high. A telescope like this was not designed for high power, high resolution work.

Not to be outdone, SkyWatcher also launched a very similar product—the Heritage 76—which also sports a tiny 3-in. mirror with a 300-mm focal length (f/4). It also comes with two, low cost eyepieces, delivering 12× and 30×, and the whole structure sits on a cute, wooden alt-azimuth mount.

The SkyWatcher Heritage 76 retails for about $10 more than the Celestron Firstscope, but is it worth the extra money? The SkyWatcher does seem more solidly built. Optically, however, there is little to choose between them. Their tiny, spherical mirrors give them limited functionality and are more suited to a beginner or a curious passerby. To get higher performance out of a grab 'n' go reflector it pays to go up in aperture.

Fig. 3.3 The SkyWatcher Heritage 76 'scope (Image © OPT. Used with permission)

299833_English

Fig. 3.4 The Skyscanner 100 reflector (Image © Orion Telescopes. Used with permission)

Orion Telescope & Binocular also launched its vision of an ultra-portable mini Dobsonian in the form of the Skyscanner 100. As its name implies, this is a 4-in. reflector, so offering significantly brighter views than the smaller 'scopes discussed above. It also sports a better figured f/4 parabolic mirror, so enabling higher power applications.

This is a *real* telescope capable of much better performance than its smaller competitors. It is arguably the ideal beginner's 'scope, collecting light to bag lots of targets that may be of interest to busy grab 'n' go astronomers. The instrument comes complete with a EZ finder and two eyepieces delivering powers of 20× and 40×, although owner reports suggest it can be pushed to powers in excess of 100×. Because field curvature is pretty prominent at f/4, you'll probably be wise to upgrade the stock eyepieces to more expensive ones that can better correct these aberrations as your experience increases. Still, this is a real bargain at $100.

Every step up in aperture brings with it the opportunity of seeing more details and more objects on your short excursions to the night sky. For reflectors, things really start to get interesting when you move above 100 mm. In this capacity, there are two 'scopes that have proven very popular with amateurs. Orion USA has successfully marketed its Starblast 4.5 portable Newtonian. Easy to carry outdoors with one hand, it is similar in design to the smaller Skyscanner 100 but with an increased aperture of 114 mm. The Starblast 114 is a much more robust instrument, though, and is a surprisingly good performer on just about any target.

Fig. 3.5 The Orion Starblast 4.5 (Image © OPT. Used with permission)

Many users report perfect collimation right out of the box. In any case, the primary mirror of each unit is centered and spotted to make adjustments that little bit easier. The Starblast 4.5 comes with an EZ red dot finder, which makes aiming the 'scope a breeze. The altitude and azimuth tension can be user adjusted to make tracking as smooth as possible. The focuser on the Starblast is a simple rack and pinion, but some users report that it can be a bit stiff. Another attractive feature of this telescope is a built-in eyepiece holder that can accommodate three 1.25-in. oculars.

The Starblast 4.5 comes with two Orion Expanse eyepieces; a 15 mm and 6 mm, with their generous 66° apparent fields, delivering powers of 30× and 75×, respectively. The f/4 mirror is well figured and will allow the user to push the magnifications to 150× or so without showing signs of breakdown—a useful power for observing the Moon, planets and double stars. Add a longer focal length wide angle eyepiece and you can enjoy breathtaking tours of the Milky Way and large deep sky objects such as the Andromeda Galaxy (M31). In summary, there's not a lot one can say against this impressive performer and it is unreservedly recommended to the canny astronomer on a budget. The tabletop Starblast 4.5 retails for about $200 but can be upgraded to a light equatorial mount for another $50.

SkyWatcher has also introduced a mid-aperture Newtonian in the form of their Heritage 130P. At its heart is a 5.2-in. f/5 Newtonian with a nicely finished parabolic mirror. Yet despite its significantly greater aperture over the Starblast it weighs about the same (6.2 kg). That's mainly attributed to the flex tube design of the Heritage, with the upper part of the tube essentially missing. It can be collapsed down into the rest of the tube for easy storage and transport. The flex tube design actually holds collimation surprisingly well.

Because the Heritage 130P is only 70 cm long, it is necessary to mount it on a garden table or some such. The all-wood Lazy Susan moves very smoothly in both azimuth and altitude, and the supplied red dot finder allows for easy pointing at celestial targets. The good quality mirror at the heart of this telescope will enable you to crank up the power on lunar and planetary targets or remain at low power as a fun, rich field telescope. The Heritage 130P can be had for the modest price of $150.

Fig. 3.6 The SkyWatcher Heritage 130P (Image © OPT. Used with permission)

A 130-mm Reflector That Finds Things

Celestron has continued to innovate in recent years, coupling good but relatively inexpensive telescopes to its go-to mounts. Some of its most successful new products is the SkyProdigy 130 reflector.

At its heart is a 5.1-in. f/5 parabolic mirror with silicon dioxide coatings to enhance the mirror's longevity. The entire telescope weights 18 pounds and can be aligned in minutes to enable the user to go to any of 4,000 objects in the night sky. The instrument comes with two eyepieces—a 25 mm and 9 mm—yielding magnifications of 26× and 72×, respectively.

Users invariably report that it's a very versatile telescope, with good optics allowing you to see satisfying views of the Moon and planets as well as the brighter deep sky objects, and all at the touch of a button. Best of all, the package can be had for as little as $699.

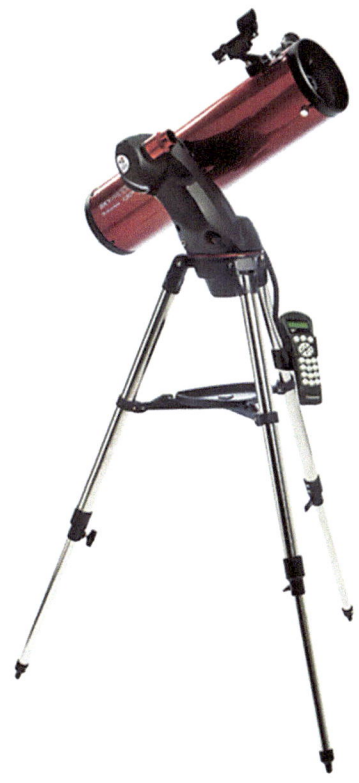

Fig. 3.7 The Celestron SkyProdigy 130 (Image © Celestron. Used with permission)

When you make the move to a 6-in. reflector—about the largest that can be considered grab 'n' go—the sky and its showpieces really start to reveal their finer details. During the 1960s and much of the 1970s, a 6-in. reflector was considered one of the best choices for a well-heeled amateur astronomer, and the same is still largely true today. They are very portable—especially when mounted as a Dob—cool off quickly and have excellent resolution and light grasp. Such a telescope has the potential to outperform even the best 4-in. refractors. Here we'll take a closer look at a few of the more popular 6-inchers on the market.

First up is yet another offering from Orion USA—the perennially popular Starblast 6, made in the same user-friendly style as its smaller sibling, the Starblast 4.5. Scott Holland, from Lowell, North Carolina, near Metro Charlotte provided his thoughts on this telescope to the author.

I have moderate light pollution at my home in Lowell, but as a member of the Charlotte Amateur Astronomy Club, I have access to a very nice dark sky site some 60 miles south in South Carolina. Over the 40+ years that I have enjoyed amateur astronomy, I have used telescopes from my first, a 30-mm spyglass to a 10-inch computerized Schmidt Cassegrain, and I've been privileged to observe through the 23-inch Alvan Clark "War of the Worlds" refractor, now located at Charles E. Daniel Observatory near Greenville, North Carolina.

When I first saw the advertisement for the Orion Starblast 4.5, I shrugged it off as just a kid's toy. After all, a 4.5 inch f/8 'scope gives respectable images, but an f/4 focal ratio demands both an excellent parabolic mirror and precise collimation to work well. For the price, I doubted that either was true. Boy, was I wrong! One night, a couple of years ago, I loaded up my old Celestron C8 Ultima, a bunch of eyepieces, charts and a red lamp and headed for our dark sky site. When I arrived, a few other club members were already there, and one of our guys had just set up his new Orion Starblast 4.5 on a modified barstool. There was also a 10-inch Dobsonian reflector and a 6-inch Maksutov set up on the club observing pad. The seeing and transparency were very good that night. Before the night was out, we were all surprised and impressed by the little Starblast's performance. If it only had a bit more aperture!

Someone else must have been thinking the same thing, because it wasn't long before ads began to appear for the new Orion Starblast 6, a 150-mm f/5 tube on a larger but very similar style mount for under $300. Screwdrivers, wrenches and pliers are usually required to assemble the mount. Here, the mount was pre-assembled. All you needed to do was to attach the tube rings to the altitude hub, open the rings, and install the optical tube. Next, the eyepiece holder would be mounted by sliding the bracket down over the screws on the braces, then tightening the screws. It couldn't have been easier.

The location of the eyepiece holder also seems to provide some additional bracing of the altitude axis. The vertical braces have cutouts to be used as carrying handles but the edges of the laminate in the cutout area can be rough on the hands if you're not careful. The altitude axis also has a rosette knob for tension adjustment located between the vertical braces. All one need do is match this tension setting to the present azimuth tension for smooth movement and tracking.

The optical tube is a standard Orion 6-inch with an f/5 primary, adjustable mirror cell, adjustable secondary, 1.25-inch rack and pinion focuser, a Vixen-style finder shoe and an Orion EZ II red dot finder. Orion sells this tube on several different mounts, including the Versa-Go as well as equatorials.

The 1.25-inch focuser has a lot of plastic in its design, but it does a decent job with the f/5 primary. The EZ II is a great finder to use with a short focus, rich-field Newtonian, especially if you're young. My only problem is bending down at the back of this short

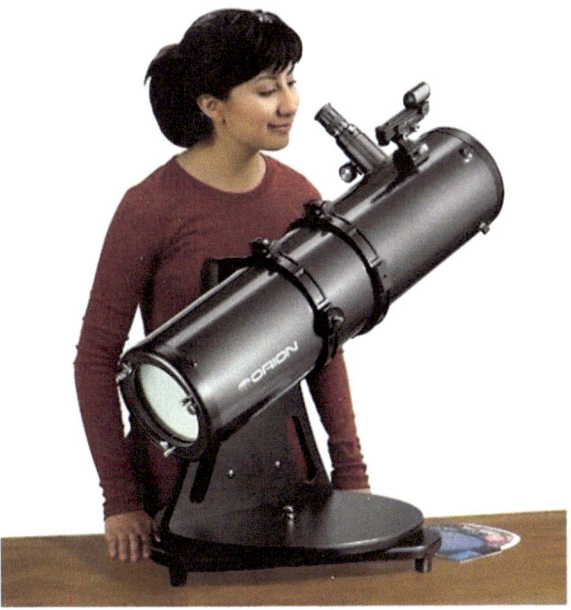

Fig. 3.8 The Orion Starblast 6 (Image © OPT. Used with permission)

telescope to get a view through the finder and to get the 'scope aimed. For an older person one should consider substituting an RACI finder in the shoe, and you're good to go. Orion provides two very nice Sirius Plossl eyepieces with the 'scope; a 25 mm and 10 mm, delivering powers of 30x and 75x, respectively. A collimation cap is also included. The primary mirror is center-marked for easy alignment of the optical train.

Overall, the Starblast 6 can be said to be an excellent grab 'n' go 'scope and an excellent value. From star testing, it seems like the optics in this 'scope are better than ¼ wave and probably more like 1/6 wave. It does great on wide field views of clusters and does a respectable job on the Moon and planets. On the negative side, it is short. You might want to put yours on a table for easy access, but with the table, vibration is a problem for a while after moving the OTA.

Orion has more recently released the new version of the Starblast 6, called the Starblast 6i. This version sells for about $100 more but is an excellent object locator, having a database of 14,000 objects.

Scott's account is fairly typical of what others have had to say about this economical mid-aperture reflector. Indeed, it's hard to see why anyone wouldn't consider it to be one heck of deal in today's market.

SkyWatcher Magic

Even a cursory survey of the amateur market will reveal that a 6-in. f/8 Dobsonian remains a popular choice among newcomers and more seasoned observers alike. In the UK, the SkyWatcher Skyliner 150P has proved especially popular.

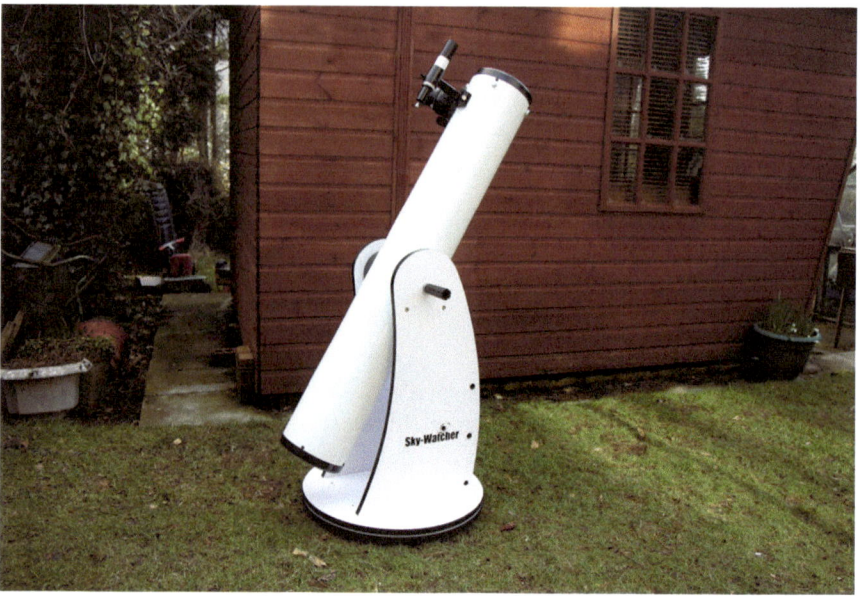

Fig. 3.9 The excellent SkyWatcher Skyliner 150P (Image by the author)

Unlike the other flex tube Dobs sold by the company, the Skyliner Ps reserve the traditional closed tube design. A new one can be had for the princely sum of £210. Arriving in two boxes, the entire telescope can be set up in less than an hour.

The Skyliner 150P is a 'no frills' 'scope and is very well made for the money. The sturdy yet lightweight rolled steel tube is finished in white. The lockable focuser is of the simple rack and pinion design lubricated by the well-derided industrial grease. It's a bit stiff, but with its generous focal ratio, achieving the best focus is relatively easy. A quick look through a homemade collimation eyepiece revealed the optics to be well aligned right out of the box. All the 1.25-in. eyepieces tested reach good focus with this telescope.

The 6×30 finder, though cheap and cheerful, makes finding things easy. The whole telescope moves smoothly in azimuth due to a well-made mount with Teflon bearings. Motions in altitude are less smooth, though, but you can still track objects manually at 200× by getting to know its quirkiness. Like some of the other models discussed, the primary mirror is center-marked for precise collimation. That's a considerably easier job to do at f/8 than at f/5 or faster. The 'scope delivers very nice views right out of the box, and a star test reveals a perfectly collimated optic. Indeed, this is generally true of all SkyWatcher Dobsonians. First impressions last. Imagine how disappointed a newcomer to the hobby would be if the mirrors were badly out of alignment, and a quick look at the Moon threw up a grotesque blur. Such an experience is potentially capable of scaring an individual right out of the hobby. Kudos to the quality control guys at SkyWatcher!

Fig. 3.10 The ever-popular Orion Skyquest XT6 (Image © OPT. Used with permission)

Of course, there are many other sources of economical 6-in. Dobsonians. As well as its Starblast 6, Orion USA has successfully marketed its longer focal length Skyquest XT6 Dob (also an f/8 focal ratio) with and without their object locator (basic package less than $300).

Other 6-in. reflectors that one could choose from include specialist dealers that usually outfit their optical tubes with better figured mirrors and superior coatings. These 'primo' dobs often have better mechanicals to enhance the viewing experience. You can read much more about these telescopes in the author's book, *Choosing and Using a Dobsonian Telescope*.

That brings us to an end of our brief overview of the grab 'n' go Newtonian market. Next, we'll take a look at the exciting new opportunities afforded by the compound telescopes, that is, those that use both lenses and mirrors to harvest starlight—the catadioptrics.

Chapter 4

Grab 'n' Go CATs

Whichever way you look at it, there really is no better time to get into visual astronomy. Many excellent and affordable telescopes are available on the market to make that a rewarding reality for newcomers eager to cut their teeth in the hobby. And that's not only true of the more traditional designs, such as refractors and Newtonian reflectors, but it is now equally valid when one considers the rich assortment of economical catadioptric telescopes (employing mirrors and lenses) that are now available to the consumer.

This wasn't always the case. Back in the 1950s, Questar Corp. designed what was arguably the most beautiful 3.5-in. (90-mm) tabletop Maksutov Cassegrain the world had ever seen.

There was no getting away from its exquisitely figured primary mirror and front corrector plate in an all-metal housing. Built-in motors effortlessly tracked the stars as the lucky owner sat comfortably in his/her garden. This telescope, which is still being manufactured today, has rightly earned its name as arguably the most stylish telescope ever conceived. But it was—and still is—extremely expensive for what it delivers at the eyepiece. A middling 4-in. refractor, for example, would show you more.

Questar enjoyed almost a complete monopoly in the catadioptric market for decades. It was just too expensive to make, but all that changed in the 1970s, when advances in electronics and product engineering saw the bringing to market of the first commercial Schmidt Cassegrain Telescopes (SCTs), first by Celestron and then by Meade Instruments.

N. English, *Grab 'n' Go Astronomy*, The Patrick Moore Practical Astronomy Series, DOI 10.1007/978-1-4939-0826-4_4, © Springer Science+Business Media New York 2014

Fig. 4.1 The Classic Questar 3.5-in. Maksutov (Image © Richard Day. Used with permission)

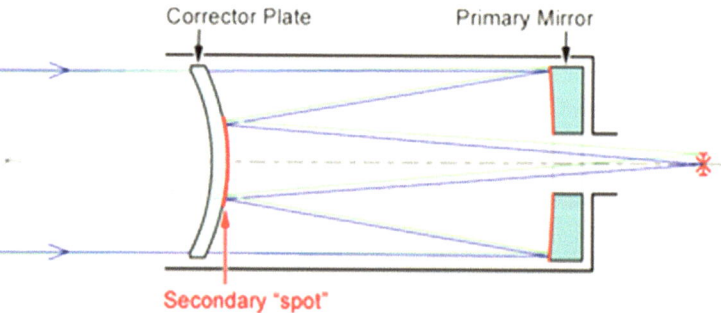

Fig. 4.2 Principle of the Maksutov Cassegrain design (Image © Oldham Optical. Used with permission)

Fig. 4.3 Principle of the Schmidt Cassegrain design (Image credit Oldham Optical)

Fig. 4.4 View of the corrector plate of a commercial SCT (Image © Celestron. Used with permission)

As you can see from the designs for both types of telescopes, they both make use of a curved glass corrector plate that directs light to the rear mirror. The light is then reflected back up the tube where it hits a secondary mirror placed on the inner surface of the corrector plate, where it is sent back down the tube a final time, where it reaches focus. When properly designed there is little to choose between Maksutovs and SCTs, but most visual observers would give the former the nod in terms of delivering finer lunar and planetary images. Because of their higher focal ratios, a given eyepiece will deliver a higher magnification and a smaller true field than an SCT of the same aperture. This might be a deciding factor for those who wish to enjoy wide field deep sky views.

The Legendary ETX

When these were first introduced by Meade Instruments Corporation in the late 1990s, they caused a huge stir among amateur astronomers. Combining excellent optics (90-mm aperture with an effective focal length of 1,250 mm (f/13.8) in a conveniently short tube, this 90-mm Maksutov was quickly and justifiably lauded as a marvel of modern technology. Sure, the mount and base were made from ABS plastic, but the images rendered by this telescope were very comparable to the legendary Questar, but offered at a price that was ten times lower! The original version—the ETX-RA—had built-in automatic tracking, but this was soon replaced by a fully computerized go-to telescope (the ETX 90 EMC) with a 30,000-object database.

Now, approaching two decades on from its inception in 1996, the ETX 90 Maksutov can still be purchased in the conventional way, that is, with full go-to capability, but can also be picked up as an optical tube assembly only, making it an exceedingly attractive spotting 'scope or grab 'n' go telescope.

The images served up by the ETX (retail price $469) are razor sharp and color free, quite like a refractor of slightly smaller aperture. Its long effective focal length means that you will get higher magnifications with your suite of eyepieces and correspondingly smaller fields of view. In this capacity it would not be a good first choice as a wide field, lower power telescope. But if you have a particular interest in the Moon and planets, the ETX will perform impressively well.

Fig. 4.5 The original ETX-RA (Image by the author)

Fig. 4.6 Portable observatory—the fully computerized Meade ETX 90 (Image © OPT. Used with permission)

Shortly after Meade launched its ground-breaking ETX 90, the company came out with two larger aperture Maksutovs: the ETX 105 and the ETX 125. These larger telescopes are essentially the original ETX on steroids. Although still excellent grab 'n' go instruments, the ETX 125 needs more time to cool off—a consequence of their closed tube design and bigger chunks of glass they possess—and thus would be on the cusp of what might be considered true grab 'n' go's.

It would be erroneous to think that Meade was the first to bring a cute, affordable little 90-mm Maksutov to market, though. That accolade went to Celestron, which in 1978 launched its 90-mm f/11 C90 Mak. Though still considered a collectible item, the optics were generally nothing to write home about. That said, although the original orange C90 has long since disappeared from the telescope stores, Celestron never gave up on it. Indeed, its modern version of the C90 (in a black tube) has received rave reviews owing to its excellent optics and super low price.

Fig. 4.7 The Celestron C90 spotter (Image © OPT. Used with permission)

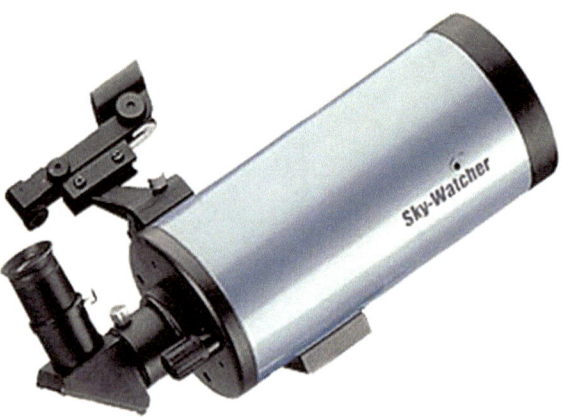

Fig. 4.8 The SkyWatcher 102 (Image © First Light Optics. Used with permission)

The C90 Maksutov comes with a decent 32-mm Plossl delivering features such as an internal flip mirror design that allows the user to view celestial objects at a 90° angle and also corrects the image orientation to view land objects at a 45° angle. The full-featured C90 Mak is also rubber covered and completely waterproof. Unlike the ETX, the C90 features durable black rubber armor for protection, a flip-up lens cap to protect the front corrector plate from dust and grime, and comes complete with its own soft-sided carrying case. Best of all is the price, though; these instruments can be had for as little as $200.

SkyWatcher has also launched its own clones of the C90 and larger Maksutovs in the form of the Skymax 90, 102 and 127. These are also available under the Orion USA re-branding (the Starmax series). These instruments can be purchased as a bare optical tube assembly or as part of a mounted package for very reasonable prizes.

User testimonies are almost universally good about these economical SkyWatcher Maks, especially for lunar and planetary work. They are easy to mount and carry about, but in the larger apertures require significantly longer to cool down to ambient temperature compared with, say, a refractor or small Newtonian reflector.

Grab 'n' Go SCTs

SCTs in all sizes have proven to be extremely popular with amateur astronomers, but only two of the smaller units can serve as versatile grab 'n' go 'scopes—the Celestron Omni XLT 127 and the C6.

Like the Maksutov Cassegrains described previously, SCTs provide sharp images that are free of false color. They lose a bit of contrast owing to their relative large central obstruction, but can provide surprisingly detailed images of all celestial targets. Because of their closed tube design, SCTs suffer from the same thermal issues during equilibration as Maksutovs when exposed to the outside air. But once these issues are dealt with, they can be deliver very satisfying views.

Weighing in at 6 pounds, the Omni XLT 127 packs a lot of punch for such a compact telescope on a stable (but not go-to) equatorial mount. It will, for example, significantly outperform a C90 on all targets, owing to its greater resolving power and light grasp.

Fig. 4.9 The Celestron Omni XLT 127 (Image © Celestron. Used with permission)

A 5-in. aperture SCT is a very good all-around performer, allowing you to see fine details in Jupiter's turbulent atmosphere, the main dark markings and polar caps on Mars and the full majesty of Saturn's ring system. To coax the best images out of this 'scope will require well cooled and collimated optics. Misalignments have been known to occur, but the instrument can be quickly readjusted for optimal performance. Deep sky views will also be decent with a telescope of this aperture. All in all, a very respectable package for $629, plus shipping.

To increase the versatility of these small SCTs, a number of companies manufacture a dedicated field flattener and focal reducer, reducing the f ratio from 10 to 6.3. Experiments carried out by this author show the images can actually be improved with this device (which easily screws into the rear cell of the instrument), reducing field curvature at the edge of the field and widening the field of view. They are also neat for astro-photographic purposes, by significantly reducing exposure times.

A Cute Addition from Vixen

Vixen Japan has introduced a nice, lightweight reflecting telescope called the VMC 110L. This is a 110-mm (4.33-in.) aperture modified Cassegrain f/9.4 optical tube assembly. The design features a small meniscus corrector lens ahead of the secondary mirror rather than the usual front corrector lens used in Maksutovs and SCTs. The advantage here is that it requires less cool-down time and is far less prone to dewing up.

It also features two 1.25-in. eyepiece holders and a flip-mirror so the view can be quickly switched from one eyepiece to another. Alternatively, a camera and eyepiece could be fitted for observing and photography.

This little Vixen catadioptric is excellent for quick excursions outside at night as well as for terrestrial observing or grab 'n' go astronomy.

Fig. 4.10 The nifty Vixen VMC 110L (Image © Vixen Japan. Used with permission)

Chapter 5

Grab 'n' Go Accessories

So you thought you had it all figured out, choosing the right telescope for your grab 'n' go adventures. Not so! It pays to consider a few accessories that will enhance your observing programs. Accordingly, this chapter will take a closer look at some other pieces of equipment that may be of benefit to you.

Are you the kind of observer that doesn't like to lug electric cables out to power your telescope but have limited time to track down the dimmer objects in your sky? If so, then why not consider the excellent hand-held object finders that have satisfied many astronomers over the last few years?

A few years back, U.S.-based companies Meade and Celestron brought out competing products that made use of global positioning systems (GPS) to produce star 'sat-navs.' The Meade version, called MySky, consists of a hand-held device that, on first glance at least, looks like a ray gun out of science fiction. It has a small screen on the back of the unit and a red illuminated gunsight-like system for aiming the device. Using an electronic compass, accelerometers and GPS sensors, this device knows its precise location on Planet Earth and where it is pointing. This means that when you aim it at a certain part of the sky, it will display a 'road map' of the stars on the rear screen and can identify the object being pointed at.

MySky will then give you information, in the form of text or audio, with the supplied earphone. It also doubles up as a finder by selecting an object from the onboard database. The Meade MySky will then display direction arrows on the screen in much the same way as a car sat-nav, which turns into a cross once you are near the intended target.

Celestron also brought out its own version of the same technology, embodied in the Sky Scout. Unlike the MySky unit, the Sky Scout prompts you to look through the device, just like you would a telescope. The circular field of view is surrounded

N. English, *Grab 'n' Go Astronomy*, The Patrick Moore Practical Astronomy Series, DOI 10.1007/978-1-4939-0826-4_5, © Springer Science+Business Media New York 2014

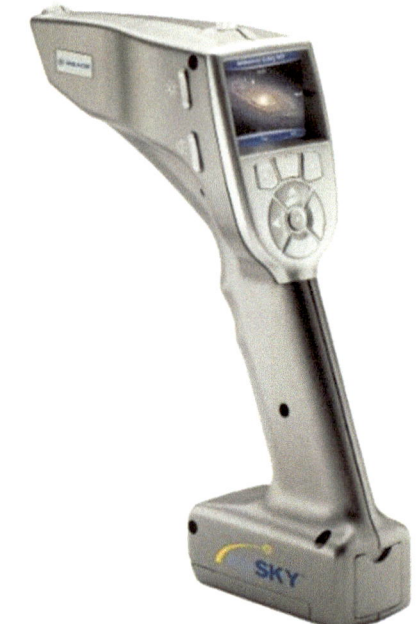

Fig. 5.1 Meade MySky (Image © OPT. Used with permission)

Fig. 5.2 Celestron Sky Scout (Image © OPT. Used with permission)

by small triangular red LED lights that act as pointers to the desired object indicated by a circle of dim red arrows, so as not to affect the user's dark adaptation.

Like Meade's offering, it can be used to both find objects to which the device is pointed to and will provide information in either text or audio format. Both these devices can be used as a high-tech finder system when used in conjunction with a telescope. It's even possible to purchase adapters to fit them to many types of telescopes or even binoculars. This makes finding or identifying objects simple for both novice and advanced observers.

These high-tech tools are often cited as being good replacements for finder telescopes when attached to a telescope. But a word of caution is prudent here. The proximity of the device to the metal tube on the telescope can interfere with the proper functioning of these devices and so can give erroneous results. They were designed to be hand-held and not used in conjunction with a telescope as a graft in go-to object locator.

All About Eyepieces

Eyepieces are to telescopes and observers as speakers are to hi-fi systems and listeners. And, like the world of sound, there are a wide variety of designs and qualities available. Choosing the right ones for your grab 'n' go observing schedule can make all the difference.

Anyone who has bought a telescope and has used an eyepiece normally supplied with it will be familiar with what it does, but what does the information normally written on the eyepiece actually tell you? For example, you receive a 25-mm Plossl with a 52° AFOV.

Let's start with eyepiece design. Originally Galileo's telescope eyepiece consisted of a single element bi-concave lens placed in the light path just in front of the focal plane of the telescope's objective. Shortly afterwards, Johannes Kepler suggested using convex lenses, improving matters somewhat by increasing the field of view significantly while rendering an upside down image. But even with these improvements, the images rendered still fell short of perfection. Over the years, several opticians attempted to produce better quality lenses and designs to improve the final view for the observer. Their goal was to produce an image free of distortions and aberrations and which magnified the image sufficiently to show detail in the object being observed. These considerations led to the introduction of improved designs like the Kellner and Huygenian.

Nowadays, we have the benefit of several hundred years of trial and error by these opticians. Over the years, though, some of the new designs have fallen out of favor as better ones were introduced. Occasionally, manufacturer's names have become interchangeable with certain designs, at least in the public's mind.

What bearing does the design of an eyepiece have on what you can see? The way lenses are shaped and how numerous they are determines the field of view, how flat it is and how well corrected the in-focus images are. The Kellner is probably one of the oldest designs still in popular use, its origins extending all the way back to 1849. Carl Kellner invented this three-element lens to reduce a common optical problem known as chromatic aberration introduced to the image. A reasonably inexpensive eyepiece to manufacture and giving a moderate field of view, it is frequently supplied with starter telescopes.

The now ubiquitous Plossl is a design that hails from 1860. The brain child of Georg Simon Plossl, it gives a still larger corrected field of view and can be manufactured in a range of useful magnifications for the modern backyard observer. Unfortunately, however, their quality can vary dramatically from manufacturer to manufacturer. In order to be well executed, a Plossl eyepiece needs to be made with high quality glasses with properly applied anti-reflection coatings, as well as blackened edges to prevent internal reflections.

Another excellent ocular that is still held in high esteem with many amateurs is the orthoscopic. Although it has a fairly narrow field of view, it shows excellent corrections for all aberrations and consequently is often still the best choice for discriminating lunar, planetary and double star observers. The more specialized monocentric eyepieces have even less air-to-glass surfaces than the orthoscopic and are sometimes recommended by diehard planetary observers as offering the sharpest.

In more recent years, there has been an increasing demand for well corrected eyepieces with extra wide fields of view. Perhaps the most successful of these wide angle eyepiece manufacturers is the Nagler line of eyepieces, named after their inventor, Al Nagler, founder of Tele Vue Optics, based in New York. Sporting a very large 82° field of view, they are made from high index glasses and multicoated to minimize stray light and maximize contrast. They are still a gold standard for eyepieces the world over.

That said, some observers have expressed a distaste for 80° class oculars, preferring instead to look through the new variety of models with 70° apparent fields of view. Arguably the finest of these were offered by Pentax in the corpus of their superlative XW line. Sadly, these have been recently discontinued, but Tele Vue Optics has taken up the gauntlet in bringing their new lines of 72° field Delos eyepieces to market. These are superlative eyepieces, combining orthoscopic-like sharpness with a wide apparent field and a very Comfortable 20-mm eye relief, enabling those who wear eyeglasses to see the entire field of view comfortably. Like most Tele Vue products, they are rather pricey, but you do get what you pay for.

Tele Vue also produce an excellent line of 100° eyepieces in the form of their Ethos eyepiece range. Many amateurs rave about the space-walk like fields served up by these hyper-wide oculars but some have expressed a dislike for the uber-expansive fields they deliver.

Another line of high performance eyepieces are encompassed by those brought to market by Explore Scientific. These Chinese-made oculars are produced to very high standards and include a range of oculars with fields of view of 70, 82, 100 and most recently an eyepiece with an incredible 120° field. Though not quite as nice those made by the market leader Tele Vue, these come impressively close at significantly lower prices to boot.

Only very rarely, though, does an experienced visual observer employ only one brand of eyepiece. Often, careful comparison between eyepieces in the field is the only way to A/B test various different models, and the outcome of such comparisons influences the individual's choice of eyepiece.

Fig. 5.3 Modern eyepieces offer well corrected, ultra wide fields, such as the new line of Tele Vue Delos series (72°) (Image by the author)

Fig. 5.4 Most amateurs own different eyepieces from different manufacturers (Image by the author)

Fig. 5.5 A standard 2× Barlow lens (Image by the author)

How many eyepieces are required? That's not an easy question to answer, but usually at least three thoughtfully considered focal lengths are necessary to cover the full range of celestial real estate one wishes to explore—low, medium and high power.

The magnification is found easily by dividing the focal length of the telescope by the focal length of the eyepiece. So, a 'scope with a focal length of 1,000 mm will give a magnification of 40× with a 25-mm eyepiece. The true field of view experienced in the eyepiece is approximated by dividing the apparent field of view by the magnification of the eyepiece. So, our 40× eyepiece with a 52° apparent field will provide a 52/40 = 1.3° true field.

In addition to a few eyepieces a good Barlow lens is also recommended. This is usually a two-element tele-negative lens that amplifies any eyepiece magnification by a factor of anywhere between 1.5× and 5×. A 2× Barlow is what is commonly found in most amateurs' eyepiece kit. Thus, three eyepieces and a good Barlow lens will deliver six different magnifications allowing, you to use your telescope for every conceivable scenario, from low power sweeps to high power planetary studies and the like. All Barlow lenses increase the eye relief—the distance between the field lens and the eye—which may or may not be desirable. One remedy, again pioneered by Tele Vue is the Powermate, which produces Barlow-like amplification without changing the eye relief of the ocular used.

The Importance of Note Keeping

Recording your observations with either written notes or a sketch is an indispensible part of the viewing experience. It can be as simple or as elaborate as you wish. Usually, the date, time and seeing conditions are noted as well as the specifications of the instrument used and the magnifications employed. This is usually followed by either a drawing of the object under study or a few words to sum up the experience, that is, a 'word picture.' You'll need a dim red light (this has minimal impact on your dark-adapted eye) to make those written recordings at the eyepiece. Note keeping will allow you to monitor any changes in the object you are studying over time and provides a framework upon which others can refer to your work, should you observe something out of the ordinary.

Astronomical Filters

For decades, eyepiece filters served as indispensible tools for the dedicated amateur astronomer. Yet, rather inexplicably, they appear to have fallen out of favor with quite a few amateur astronomers of late. That's a great pity, as exploiting their properties—some of which are quite remarkable—judiciously will only serve to enhance your observing program.

What do filters do? Simply put, they can improve what we see by removing what we don't want to see from the view. Think of Polaroid glasses on a bright day! We can all understand that most contemporary observers desire the most aesthetically pleasing image possible from a 'scope. Adding a color filter won't do anything to enhance that world view, but what it will most definitely do is exaggerate differences in brightness between the various features of a planetary or lunar image.

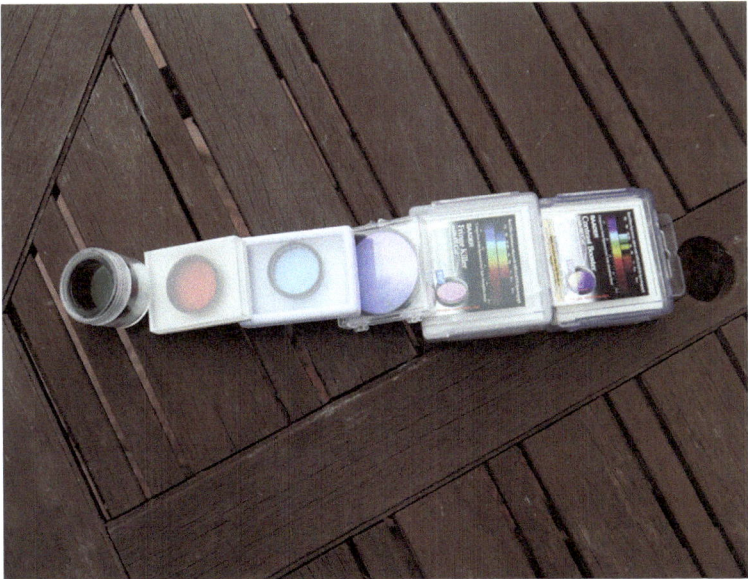

Fig. 5.6 Filters can improve your telescopic views (Image by the author)

Color Filters

Color filters (the properties of which are listed in Table 5.1) have many uses, including:

- Glare reduction, which almost invariably leads to an increase in perceived image quality.
- Overcoming, to a greater or lesser degree, the image-distorting effects of the atmosphere.
- Enabling observers to study different levels of a planetary atmosphere.
- Increasing contrast between areas of different color.
- While not eliminating optical defects, improving image definition even with bad or mediocre optics.

Most of the attributes of filters highlighted above are well known, with the possible exception being the second one on the list. Meteorologists have known for quite some time about the scattering effects of particles in the atmosphere. Known as Rayleigh scattering, it predicts that for a given-sized particle, light is scattered in inverse proportion to the fourth power of wavelength. Thus, it can be shown that violet light (wavelength ~400 nm) is scattered some 16 times more effectively than deep red light (800 nm). That's why the sky is blue and sunsets are red.

Thus, so the theory goes, employing a red filter during turbulent atmospheric episodes might mitigate, to some degree, the deleterious effects of bad seeing. A violet (47A) filter is very useful for observing cloud features on Venus, and although its light transmission is painfully low, it can be pressed into service with larger aperture 'scopes.

Mars is a great planet to learn how good color filters can be in extracting atmospheric and surface features. A simple light yellow (#8) reduces glare and increases

Table 5.1 Transmission data for various Wratten color filters

Color filter	Wratten #	Light transmission (%)
Light yellow	8	83
Yellow-green	11	78
Yellow	12	74
Deep yellow	15	67
Orange	21	46
Light red	23A	25
Red	25A	14
Dark blue	38A	17
Violet	4	3
Light green	56	53
Green	58	24
Blue	80A	30
Light blue	82A	73

contrast in smaller apertures (5 in. and less). An orange (#21) is useful for penetrating haze and cloud in the Martian disk, as well as increasing contrast between
the light and dark areas of the planet. A light green #56 filter darkens both red and
blue features, enabling the observer to prize the morphology of the polar cap more
easily.

Jupiter and Saturn also benefit from colored filtration. Blue and green ones are
just dandy for bringing out the belts of the planets. A yellow filter can help reveal
bluish features (festoons), while a red filter can help bring out the white ovals so
cherished by planetary observers. The icy Saturnian ring system, too, can look
majestic using a red filter.

Although older, dyed-in-the-glass filters are perfectly adequate (and cheap as
chips). One might gain some additional benefit from the newer interference color
filters manufactured by companies such as Baader Planetarium in Germany. These
have multiple layers of dielectric coatings that are finely polished and guaranteed
to be optically flat to 1/4 wave across their entire surface. This means that can be
placed well ahead of the eyepiece in the optical train without causing any significant degradation of the image.

A new breed of interference-based filters have become available to the amateur
astronomer. As their name suggests, these filters act by selectively blocking the
contrast-robbing effects of unfocused light in achromatic images. Perhaps the most
versatile are those produced by Baader Planetarium, which manufacture the Fringe
Killer, Semi-Apo and Contrast Booster. A study of these filters conducted by the
author on a number of economical achromatic refractors has shown that they can
be extremely effective at cleaning up an image, especially on bright objects such as
planets. They help the user find the best focus, reducing glare and sharpening the
image. They also work well with apochromatic refractors as well as Newtonians
and SCTs. All, in all, very useful tools for the dedicated visual observer.

Improving Resolving Power

It is also possible to improve resolving power with color filters. The resolving
power of a telescope (in radians) is approximated by Lambda/D, where Lambda is
the wavelength and D is telescope aperture. The Dawes limit is closely matched to
a wavelength of 562 nm. Converting radians to angular degrees, we can easily
compute that for a 4-in. instrument (0.1 m), the Dawes limit is ~1.15 arc seconds.
Yet, there are quite a few instances where this value has been exceeded. Famously,
Dawes himself appears to have broken his own empirical formula. Indeed, this
author has obtained evidence that many observers employing large classical refractors have exceeded the Dawes limit in their measures of double stars.

Although all of this sounds like pie in the sky, it can be handily demonstrated
with color filters. A violet filter working at 390 nm will improve the resolution of
a telescope by up to 30 %! Noted CCD imager Damian Peach produced a cool
illustration of this effect. You will note that the binary system is unresolved at red
wavelengths, elongated at green wavelengths, and cleanly resolved at blue (shorter)
wavelengths.

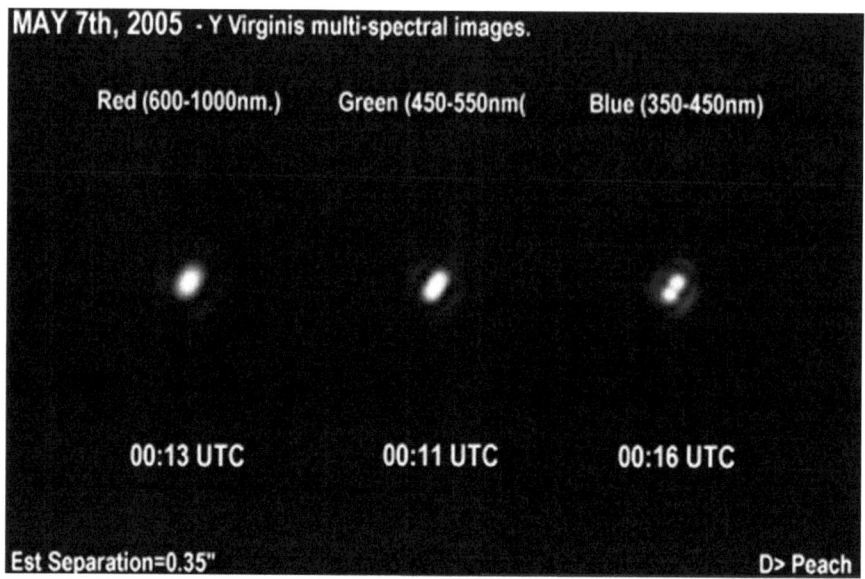

MAY 7th, 2005 - Y Virginis multi-spectral images.

Red (600-1000nm.)   Green (450-550nm(Blue (350-450nm)

00:13 UTC   00:11 UTC   00:16 UTC

Est Separation=0.35"   D> Peach

Fig. 5.7   The effects of color filters on resolving power. Each image was shot through the same telescope but at different wavebands (Image credit and © Damian Peach)

Polarizing Filters

The light that reaches us from the depths of space vibrates in every conceivable plane. Plane polarized light, on the other hand, vibrates in only one plane, greatly reducing scattered light in the eye (irradiance) and increasing contrast. Variable polarizers have been around for quite some time, but, with a maximum transmission of 40 perfect, were restricted to use in large aperture instruments. Recently, however, Bill Burgess, founder of Burgess Optical, has developed a much more versatile dielectric polarizing filter. Called the Planetary Detail Enhancing System (PDES), it consists of a pair of continuously variable polarizers that can transmit between 5 and 90 % of the light passing through it.

After screwing the filter into the eyepiece, it then slots into a sleeve. Two nylon screws are then tightened to add just enough friction to enable the observer to vary the degree of illumination simply by rotating the eyepiece.

This filter was put through its paces using both 5- and 6-in. refractors (127 and 152 mm, respectively) over a period of several weeks and was found to be quite impressive. As well as reducing glare, it also improved the images of shorter focal length achromatic refractors. Studying Jupiter and Saturn with this filter in place, you will quickly form the impression that you are seeing more atmospheric detail without causing a color shift to the image. It's also a most excellent tool for reducing glare in tricky double stars where one component is much brighter than the other. Although more expensive than simple color filters, the Burgess PDES (avail-

Fig. 5.8 The Burgess Optical PDES kit (Image by the author)

able from www.burgessoptical.com) is a new and exciting tool for the dedicated visual observer.

Spread the word: color, polarizing and minus violet filters are useful tools. Find the time to use them skillfully.

Nebula Filters

Many amateur astronomers sing the praises of light pollution and nebula filters to bring out details of many faint, deep sky objects. Nebula filters transmit only a narrow band of spectral lines from the spectrum (usually 22 nm or less). As their name implies, these are mainly used for nebulae observation. Emission nebulae mainly radiate in the light of doubly ionized oxygen in the visible spectrum, which emits near the 500 nm wavelength. These nebulae also radiate weaker at 486 nm from the hydrogen-beta atoms.

There are three main types of narrowband filters for visual use: Ultra-High Contrast (UHC), Oxygen(O)-III and Hydrogen-Beta, the narrowest of the three filters with an 8 nm range. The UHC filters range from 484 to 506 nm. They transmit both the O-III and H-Beta spectral lines, blocking a large fraction of light pollution and bring out the details of planetary nebulae and most of emission nebulae under a dark sky, as well as darkening the background sky. If this author were to choose one, it would be the UHC filter as it provides an excellent compromise between.

Dew Removal Systems

For the majority of amateur astronomers, the onset of dew is a real threat to making the most of an extended period of observing, with only those living in low humidity climes being free from it. The simplest defense against the buildup of dew on your optics is a dew shield, which usually accompanies most refractor telescopes. For catadioptric and Newtonian telescopes, a wide variety of low-cost flexi-dew shields can be purchased to delay the onset of dew. And while these certainly retard the formation of dew on your optics they are not a permanent panacea.

There are two active ways of preventing dew formation. One way involves using a simple electric hair dryer that blows off the condensed moisture on your optics. Although this method certainly works, it delivers a little too much heat to the lens or mirror, making the image look distorted for a short while. Moreover, the dew will return as soon as the optics cool down again, thus necessitating the application of several blasts of hot air over the course of a short time.

A much superior way of controlling the formation of dew is to acquire a good dew removal system. The simplest systems usually take the form of a string of resistors arranged inside a tube that dissipates electrical heat at just the right rate to stop the formation of dew in the first place but not in such a way to overheat the optics. These can be purchased inexpensively or indeed constructed from easy to obtain parts from the Internet. Other companies, such as Kendrick Astro Instruments, offer more sophisticated dew removal systems for state of the art prevention of dew formation.

Chapter 6

Choosing a Mount for Your Telescope

It has to be said. A telescope is only as good as the mount it rides on. And although many of the Newtonians and catadioptrics we have discussed in earlier chapters come with their own dedicated mounts, small refractors—which have proven to be enormously popular as grab 'n' go telescopes—often come as an optical tube assembly only. Telescope mounts come in a bewildering number of varieties to suit everyone's budget. Since we are confining ourselves in this book to small, grab 'n' go telescopes, we shall only consider those lighter weight mounts currently available on the market.

Broadly speaking, mounts can be placed into two categories; alt-azimuth and equatorial. First we will discuss the kinds of alt-az mounts that have sprung up over the last decade.

The SkyWatcher AZ3 Mount

Perhaps the simplest alt-azimuth mount currently available to amateur astronomers is SkyWatcher's AZ3 mount. This is a very straightforward alt-azimuth mount with an adjustable aluminum tripod and a large accessory tray. It features two flexible slow-motion cables, one for vertical and one for horizontal control. One can accurately track celestial objects with ease. This mount is ideal for terrestrial observations but can also be used for a no-fuss grab 'n' go astronomy setup. This is a great mount for small, short tube refractors up to 80 mm in aperture. It is also adequate for small Maksutov Cassegrains and Newtonians up to about 4 in. (102 mm) in aperture.

N. English, *Grab 'n' Go Astronomy*, The Patrick Moore Practical Astronomy Series, DOI 10.1007/978-1-4939-0826-4_6, © Springer Science+Business Media New York 2014

Fig. 6.1 The SkyWatcher AZ3 mount (Image © SkyWatcher USA. Used with permission)

Sturdier still are the Vixen Porta mounts. The regular version of this mount handles telescopes up to 3.5 kg in weight. One can move the telescope either by manually pushing it or using slow-motion handles and its built-in friction control system. The instrument is simply attached via the mount's 2-in. dovetail bracket. A two-section lightweight aluminum tripod can be raised to 1.8 m above the ground. The heavier version of the same mount—the Vixen Porta II—is essentially a more massive clone of the original Porta, allowing telescopes up to twice the weight to be used profitably.

If you already have a sturdy photographic tripod, then it's possible to retrofit it with an alt-azimuth head that can carry a small telescope. One example is offered by Stellarvue, USA. Called the M2D ($279), this mates directly to a standard camera tripod with a 3/8-16 bolt and can accommodate telescopes up to 20 pounds in weight.

Those requiring a step up in stability would do well to consider one of a variety of new heavy duty alt-azimuth heads that can hold either a single or two larger optical tubes. Take, for example, the Vixen Castor mount ($299). The alt-azimuth head attaches directly to photo tripods using the standard 3/8-16 bolt.

Altair Astro, UK, have recently produced their own alt-az mount for their impressive range of small and medium aperture telescopes. Called the Sabre, the unit can be retrofitted onto standard EQ5 tripods to handle a wide variety of grab 'n' go telescopes. This author has had the pleasure of putting the Altair Astro Sabre

Fig. 6.2 The Vixen Porta II mount (Image by the author)

Fig. 6.3 The Stellarvue M2D (Image © Stellarvue. Used with permission)

Fig. 6.4 The Vixen Castor alt-az mount head (Image © OPT. Used with permission)

Fig. 6.5 The Altair Astro Sabre mount (Image by the author)

through its paces and found it could take telescopes as big as 30 pounds without flinching. Although it doesn't possess slow motion controls, it does sport beautifully machined adjustable tension knobs on both axis to ensure the perfect amount of traction while moving the telescope manually across the sky.

For those requiring the most stable of alt-azimuth mountings on a budget, you need look no further than something like the SkyWatcher SkyTee II alt-azimuth mount. It has great versatility. Like the Vixen Castor, the SkyTee can accommodate two telescopes, but its load capacity is substantially greater than the former (15 kg).

Fig. 6.6 The SkyWatcher SkyTee II alt-azimuth mount (Image by the author)

The manual slow-motion controls are satisfyingly smooth, allowing the user to track a celestial object at high powers with confidence. A pier can be attached to elevate the telescope as much as possible above the ground. It works well in cold weather too, where many other mounts would begin to struggle. All in all; a very versatile platform for a modest investment!

Another exciting range of alt-az mounts to hit the market are provided by the U.S.-based company DiscMounts. These were designed with a patented variable friction control along with a large friction disc. This allows the mount to be adjusted to accommodate large changes in eyepiece (or camera) weights. All that is needed is to adjust the mount to accommodate for the telescope configuration.

These mounts are very compact and look like a solid cube. What sets them apart from the competition is their rigidity (at least almost), having a vibration damping material between the discs. In addition the telescope is mounted to the side of the mount. This allows the telescope to be mounted to the altitude axis disc (with proper rings, etc.) and also allows it to rotate on both axes 360° (depending on tri-pod legs, etc.). The lightest DiscMount—the DM4—has a load capacity of 18 pounds, while the bulkier DM6 can carry optical tubes up to 40 pounds in weight. As with everything else in this market, you have to pay for such quality. The DM4 head (that is, without a tripod) retails for $800, while the DM6 head can be had for $1,200. DiscMounts also offer an extensive list of accessories to maximize the performance of their higher quality alt-az mounts.

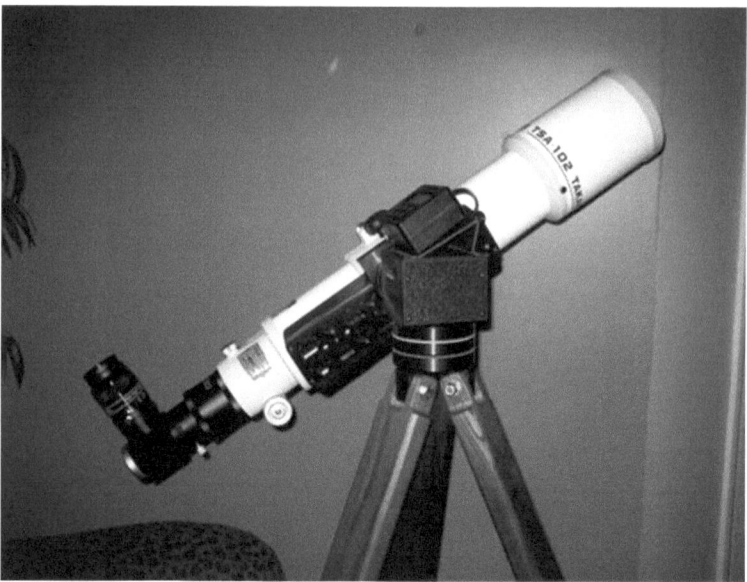

Fig. 6.7 A 4-in. refractor astride a DM4 from DiscMounts (Image © DiscMounts. Used with permission)

Equatorial Mounts

Unlike alt-azimuth mounts, which require the user to move the telescope both in azimuth and altitude to keep track with a celestial target, an equatorial mount, once aligned on the celestial pole, enables the telescopist to follow an object with only one adjustment. Coupled to a motorized mount, equatorials can be exceptionally easy to use and let you spend more time observing rather than chasing your target.

Fortunately, many lightweight equatorial mounts are available to the grab 'n' go astronomer. One of the best used and well thought of is the SkyWatcher EQ5, which can be set up in just a few minutes and allows you to mount a sizeable telescope for celestial viewing.

The sturdy EQ5 equatorial mount provides good precision and stability and a host of other useful features for the serious astronomer. The mount rests firmly on a large, adjustable stainless steel pipe tripod with accessory tray. It is supplied complete with a bubble level, latitude adjuster with micrometer scale, as well as an azimuth polar-alignment adjuster. It also features engraved aluminum setting circles and manual slow-motion tracking controls. An optional polar 'scope is available for aiding in polar alignment. Optional single-axis and dual-axis D.C. motor drives can be installed for auto-tracking purposes.

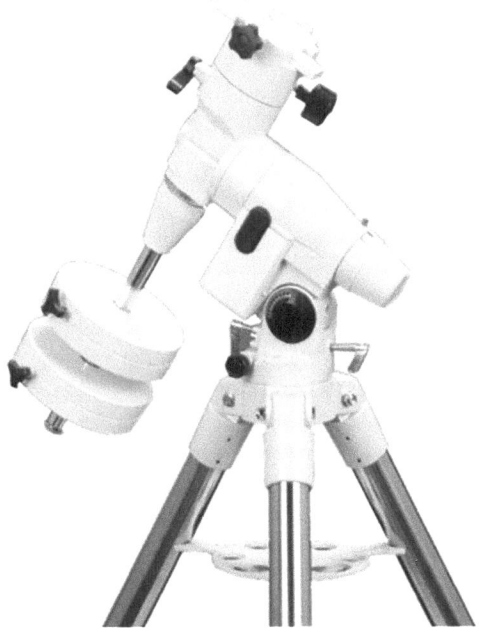

Fig. 6.8 The tried and trusted SkyWatcher EQ5 equatorial mount (Image © FirstLight Optics. Used with permission)

If you own a very small telescope, such as an 80-mm short tube refractor, then SkyWatcher's lighter weight EQ3-2 mount may be all you need. The EQ3-2 mount has received high praise from experienced observers. The mount rests on a large, adjustable aluminum tripod with accessory tray. A bubble level allows for quick and accurate leveling. Slow-motion control knobs allow for continuous manual tracking of celestial objects. The precision mount features full 360° worm-gear tracking controls on both R.A. and DEC. axes (motors purchased optionally). Dual metal setting circle dials allow quick location of the target using celestial coordinates. An optional polar 'scope is available for aiding in polar alignment. Optional single-axis and dual-axis D.C. motor drives can be installed for auto-tracking purposes.

These type mounts are sold by other manufacturers, Vixen and Celestron, as well as the American re-branders, Orion Telescope & Binocular. And although the various units look much the same, you generally get what you pay for.

Go-To Mounts

Many grab 'n' go telescopes either come with or can be coupled to computerized go-to mounts. These have built-in GPS technology that allow them to determine their precise location on Earth and once activated can accurately locate the positions of any one of thousands of celestial objects that are above the horizon at that time. Suffice it to say that go-to technology has revolutionized the way we observe as amateur astronomers.

Go-to capability is, of course, nothing new. This new technology goes back at least to a line of computerized Celestron and Meade telescopes that were first marketed in the 1990s. Over the years since these pioneering new products emerged, go-to has been offered on an ever increasing number of telescopes, with the result that nowadays it is not unusual to find it on budget packages.

We've already covered a few go-to models in previous chapters, such as the Meade ETX and Celestron Nexstar series. Originally, these telescopes could only be purchased as part of a package that included the mount. These days, one can purchase go-to mounts separately. Take, for example, the Celestron NX SE 6/8 available from good retailers for as little as $380. This computerized alt-azimuth mounts have NexStar technology making them fully go-to capable. The included hand controller contains a 40,000-object database and offers a variety of different tracking and slew options. With the included NexRemote and Sky X First Light

Fig. 6.9 The Celestron NX SE 6/8 (Image © OPT. Used with permission)

Fig. 6.10 The iOptron CubePro GoTo alt-az mount (Image © OPT. Used with permission)

software, plus the RS232 cable you have everything you need to control the mount with your computer if you so choose. A sturdy steel tripod ensures a stable footing, and the mount's built-in saddle plate accepts V series dovetails that makes attaching your telescope easy!

The Celestron NX SE 6/8 would be a good match for any telescope under about 12 pounds.

Another popular go-to mount is the iOptron CubePro GoTo altazimuth mount. It's the ideal portable mount to go with your portable telescope. The CubePro features a SmartStar computerized control system with 130,000 objects and an 8-line backlit LCD screen. Featuring a 32-channel internal GPS, its easy alignment procedure and accurate go-to and auto-tracking minimize the setup time. The 1-in. diameter stainless steel tripod legs with metal platform and metal hinges makes the mount very sturdy. The compact design of the tripod and the mount makes it easy to carry around in a travel case, and the mount can be fully assembled in under five minutes. The standard dovetail makes this mount compatible with many different optical tube assemblies of 12 pounds and under. In addition, the CubePro is compatible with all ASCOM-compliant planetarium programs and many others, such as *Sky X, Starry Night* and *Voyager*.

Fig. 6.11 The Meade LXD 75 equatorial mount (Image © Meade.com. Used with permission)

The aforementioned models are just a few of the wide range of mounts available to today's amateur. As with everything else, it pays to look around to see which product suits your particular needs in grab 'n' go mode. Good luck with your purchase!

In addition to the aforementioned alt-azimuth go-to mounts, a number of reasonably lightweight equatorial mounts with go-to capability have been around for over a decade now. Here we'll mention only two—the Meade LXD 75 mount and the Vixen SXD.

In essence a clone of the widely regarded Vixen Great Polaris (GP) mount, the Meade LXD55 can comfortably accommodate a 5-in. refractor, 6-in. Newtonian or 8-in. SCT. Using the supplied hand controller, the user can slew the telescope to any of 30,223 celestial objects and have the telescope both center and track the object in the field of view. Although reports from amateurs in the field are generally favorable, some have resorted to completely stripping down the mount and re-lubricating with better quality grease and more durable metal parts.

The Vixen SXD mount can, in many ways, be considered to be a more sophisticated version of the Meade LXD 75. The innovative STAR BOOK hand controller offers astronomical navigation with a large full-color LCD screen.

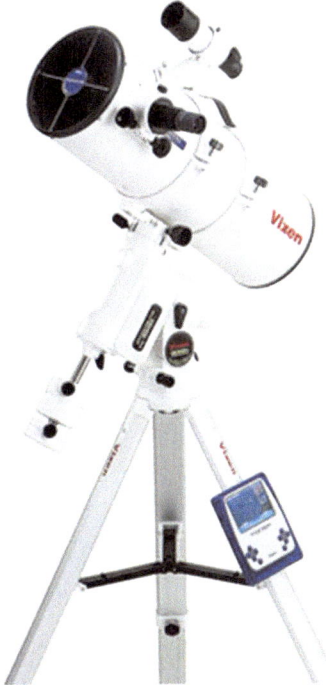

Fig. 6.12 The Vixen SXD equatorial go-to mount (Image © Vixen Optics. Used with permission)

The color screen is a detailed star chart that acts as a navigational and go-to guide for the mount. The database contains over 22,000 objects, and others can be added. Zoom-In/Out buttons for the color star chart display also control the motor speed. As you zoom in closer to an object on the display screen, the motor speed automatically becomes finer, and as you zoom out the motor speed increases to cover larger distances.

Part II

Grab 'n' Go
Objects
for Viewing

Chapter 7

Our Stellar Universe

Congratulations! You've successfully navigated your way through the bewildering number of products available to the budding grab 'n' go astronomer. You've chosen your telescope, its accessories and mount and are ready to explore the universe from the comfort of your backyard or your vacation dark sky site. In this part of the book we'll be exploring the rich milieu of celestial real estate ready for observing. The subjects discussed are rather selective, but they will serve to give the reader a flavor of the kinds of things that can be profitably observed with a small telescope.

In order to use this section of the book effectively, it's important to make use of a good star atlas to locate the pantheon of objects that we shall visit as we move through its various chapters. To this end, a short list of recommended atlases is referenced in the bibliography at the back of this book.

The telescope used to study these objects is a trusty Stellarvue 80/9D achromatic refractor. The objective lens has an aperture of 80 mm and a focal length of 750 mm (f/9). Its excellent optics can be used to obtain low power, wide-angle views of star fields as well as high power applications where magnifications in excess of 200× can be fruitfully employed.

So, without further ado, let's get going.

N. English, *Grab 'n' Go Astronomy*, The Patrick Moore Practical Astronomy Series, DOI 10.1007/978-1-4939-0826-4_7, © Springer Science+Business Media New York 2014

Fig. 7.1 The Stellarvue 80/9D achromatic refractor (Image by the author)

A Journey Through the Spectral Sequence

In our first grab 'n' go adventure, we'll explore the colors of the stars and the veritable mine of information they provide us in understanding their evolution through birth, life and death.

In astronomy, stellar classification involves grouping stars together based on their spectral characteristics. The *spectral class* of a star gives clues to the temperature of its atmosphere and an objective measure of the photosphere's temperature. Light from the star is analyzed by splitting it with a diffraction grating, subdividing the incoming photons into a spectrum exhibiting a rainbow of colors interspersed with absorption lines. Each line represents the unique signature of a certain chemical element or its derivatives (normally the positively charged ions of those elements).

The presence of a certain chemical element in such an absorption spectrum primarily indicates that the temperatures are suitable for a certain excitation of this element. If the star temperature has been determined by a majority of absorption lines, unusual absences or strengths of lines for a certain element may indicate an unusual underlying chemical composition.

Most stars are currently classified using the letters O, B, A, F, G, K and M, where O stars are the hottest and the letter sequence indicates successively cooler stars up to the coolest M class. Handy mnemonics for remembering the spectral type letters are "Oh Be A Fine Girl Kiss Me" or "Oh Boy An F Grade Kills Me." According to informal tradition that has stood the test of time, O stars are often referred to as "blue," B stars are called "blue–white," A stars are called "white," F stars are called "yellow–white," G stars are called "yellow," K stars are called "orange," and M stars, "red," even though the actual star colors noted by an observer may deviate from these colors, depending on visual conditions and individual stars observed.

The current non-alphabetical scheme developed from an earlier scheme using all letters from A to O; the original letters were retained, but the star classes were re-ordered in the current temperature order when the connection between the stars' class and temperatures became clear. In addition, a few star classes were dropped as duplicates of others.

In the current star classification scheme, the so-called *Morgan-Keenan system*, the spectrum letter is further subdivided by a number system ranging from 0 to 9, indicating tenths of the range between two star classes. So, for example, an A5 star is half way (five tenths) between A0 and F0. Lower numbered stars in the same class are hotter. Another dimension that is included in the Morgan-Keenan system is the luminosity class, expressed by the Roman numbers I–V, referring to the width of certain absorption lines in the star's spectrum. It can be demonstrated that this feature is a general measure of the size of the star, and thus of the total luminosity output from the star. Luminosity class I stars are *supergiants*, class II are bright giants, class III ordinary *giants* and class V either *dwarfs* or more properly *main-sequence stars*. For example, the Sun has the spectral type G2V, which might be interpreted as "a 'yellow' two tenths towards 'orange' main-sequence star." In contrast, the brightest star in the sky, Sirius, is designated A1V.

It's important to stress that the mass, radius and luminosity listed for each class are appropriate only for stars on the main-sequence portion of their lives (the most stable and long-lived period in a star's life where hydrogen is being fused into helium in its core) and so are not appropriate for more evolved red giants or less highly evolved protostars.

The reason for the odd arrangement of letters is entirely historical. An early classification of spectra by the Italian astronomer, Father Angelo Secchi, in the 1860s divided stars into those with prominent lines from the hydrogen Balmer series (group I, with a subtype representing many of the stars in Orion); those with spectra which, like the Sun, showed calcium and sodium lines (group II); colored stars, whose spectra showed wide bands (group III); and carbon stars (group IV).

In the 1880s, the astronomer Edward C. Pickering began to make a survey of stellar spectra at the Harvard College Observatory. A first result of this work was the *Draper Catalogue of Stellar Spectra*, published in 1890. Williamina Fleming classified most of the spectra in this catalog. It used a scheme in which the previously used Secchi classes (I–IV) were divided into more specific classes, given letters from A to N. Also, the letters O, P and Q were used, O for stars whose spectra consisted mainly of bright lines, P for planetary nebulae, and Q for stars not fitting into any other class.

In 1897 another worker at Harvard, Antonia Maury, moved the Orion subtype of Secchi class I ahead of the remainder of Secchi class I, thus placing the modern type B ahead of the modern type A. She was the first to do so, although she did not use lettered spectral types but rather a series of 22 types numbered from I to XXII. In 1901, Annie Jump Cannon returned to the lettered types, but dropped all letters except O, B, A, F, G, K and M, used in that order, as well as P for planetary nebulae(explained later) and Q for some peculiar spectra. She also used types such as B5A for stars halfway between types B and A, F2G for stars one-fifth of the way from F to G, and so forth. Finally, by 1912, Cannon had changed the types B, A, B5A, F2G, etc. to B0, A0, B5, F2, etc., classification system. And the rest, as they say, is history.

A Word About Stellar Magnitudes

The first thing that strikes any visual observer who gazes at the sky is that the stars appear to have different brightness levels, or *apparent magnitudes*. The qualification 'apparent' implies that this is how bright the star apparently looks without taking into account its true brightness. Like most other systems in astronomy, the stellar magnitude system evolved over many centuries but suffice it to say here that a star with a visual magnitude of +1 will be about 2.512 times brighter than another with a magnitude of +2. Conveniently stars separated by six visual magnitudes vary in brightness by a factor of 100.

In urban and other heavily light polluted areas, the typical naked eye limit is about +5 but in exceptionally dark and transparent skies, the most acute observers can detect stars down to about +8.

Of course, the magnitude brightness scale tells us next to nothing about whether a star is bright because of its proximity to the solar system or whether it is due to its greater intrinsic luminosity. To that end, astronomers define an *absolute magnitude* to the stellar bodies. This is defined as the brightness an object would have at a distance of ten parsecs from us (one parsec is 3.26 light years).

Activity: Exploring the Spectra

Go out on the next winter evening and look for the following easily found stars. Examine each one at low and medium power, noting carefully the color of the star as it appears to your eyes.

Star	Spectral class
Theta C Orionis	O6
Beta Cassiopeaie	B1
Vega	A0
Sirius	A1
Alpha Persei	F5
Capella	G8
Aldebaran	K5
Betelegeuse	M1

Once you have made a mental note of their colors, select another dozen stars randomly across the sky and make a note of their colors. Can you guess the spectral class of the star from its color? With practice, you'll become quite skilled at doing this.

The Hertzsprung–Russell (H–R) Diagram and the Concept of Stellar Evolution

Our Sun formed about 4,600 million years ago and is now about halfway through its life. Diligent research has shown that during this epoch the levels of the radioactive elements—uranium and thorium—essential for building planets with long-lived plate tectonics, were at their highest. Moreover, there is now evidence that the solar nebula, that great cloud of gas and dust out of which the Sun and its retinue of planetary siblings formed—was enriched by an eclectic mix of elements generated from not one but two different types of stellar explosions (supernovae events). These exploding stars were not so close as to destroy the solar nebula but neither were they so far way as to make no difference to its chemical constitution. How uncanny!

Scientists have discovered many different kinds of stars—from dwarfs to supergiants, and from the main sequence to old, giant, pulsating stars called Cepheids. But up until the early 1900s, there was no general way to classify them. All that changed with the invention of the Hertzsprung–Russell (H–R) diagram, which has become one of the most important tools in stellar astronomy. You could say that the H-R diagram is to astrophysicists what the Periodic Table of Elements is to chemists.

Independently, Danish astronomer Ejnar Hertzsprung and American astronomer Henry Norris Russell discovered that when they compared the luminosity with the type of light that was observed from stars, there were many patterns that emerged. In 1905, Hertzsprung presented tables of luminosities and star colors, noting many correlations and trends. In 1913, Russell published similar data in a diagram. It is now called the Hertzsprung–Russell diagram in honor of these two pioneers.

Russell noticed that almost 90 % of the stars fell along a diagonal ribbon that stretched from the top left to the bottom right of his diagram. The stars that fell into this diagonal ribbon were classified as being on the *main sequence*, of which our Sun is a member.

They also noticed that other groups of stars became evident. Those included blue supergiants in the upper left, red supergiants in the upper right, white dwarfs in the lower left, red dwarfs in the lower right, red giants and Cepheid variables in the middle-right (the branch extending from the main sequence). These stars will be discussed in more detail in due course.

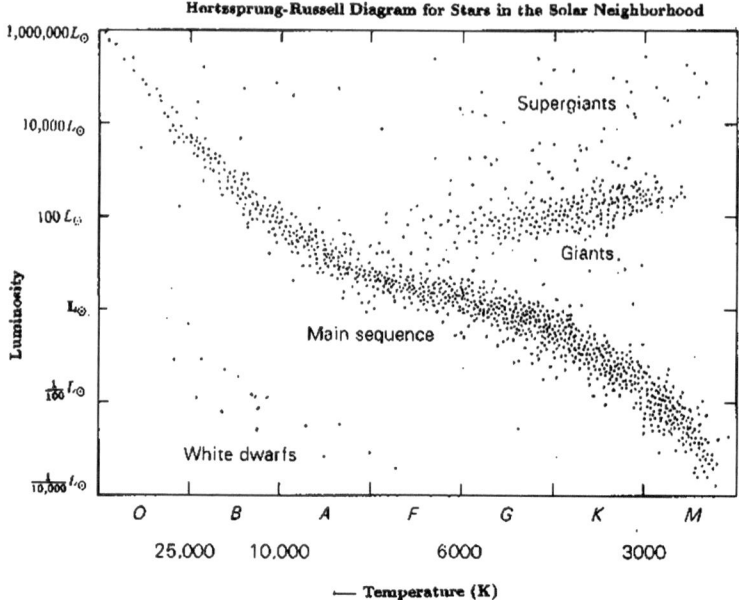

O, B and A stars are sometimes called "early type," while K and M stars are said to be "late type." This stems from an early twentieth-century model of stellar evolution in which stars were powered by gravitational contraction via the Kelvin-Helmholtz mechanism in which stars start their lives as very hot "early-type" stars, and then gradually cool down, thereby evolving into "late-type" stars. This mechanism provided ages of the Sun that were much smaller (of the order of a few tens of millions of years) than what is observed, and was rendered obsolete by the discovery that stars are powered by nuclear fusion. However, the smallest stars, the brown dwarfs, whose energy comes from gravitational contraction alone, cool as they age and so progress to later spectral types. For example, the highest-mass brown dwarfs start their lives with M-type spectra and will cool through the L, T and Y spectral classes.

Understanding the Spectral Sequence

In continuing our visual study of the spectral classification scheme, we now turn our attention to the brightest luminary in the constellation of Auriga—the lovely yellow **Capella,** shining across 42 light years of space. Visible in the eastern sky during Winter, a telescope clearly reveals the star to be yellowish white, certainly

Fig. 7.3 Image courtesy of NASA

consistent with its G8 spectral classification. No companions are seen in an 18-mm orthoscopic (yielding 42×), nor indeed at any greater enlargements, but we know that Capella has a close companion, another yellow giant orbiting every 104 days. The companion is betrayed by studying the star spectroscopically, i.e., it is a *spectroscopic binary.*

Although Capella's G8 designation might suggest a star happily enjoying life on the main sequence, astronomers now know that it has in fact left it, i.e., its core hydrogen burning days are over. Capella is a more highly evolved star than our Sun, slowly transforming itself into a red giant. It will continue to expand and cool, shedding much of its mass in a series of shells enveloping the core. A torrent of high energy waves—mostly ultraviolet and soft X-rays—will light up these shells like a neon bulb, creating what we refer to as a *planetary nebula.* And what's left behind, the cinders of a modest-sized star, will 1 day become a *white dwarf.* White dwarfs are death stars that litter the bottom of the HR diagram, under the main sequence. The Sun, in all its glory, is fated to end up the same way in the far distant future.

So how do stars continue to shine once they leave the main sequence? As soon as the star has used up its reserves of hydrogen in its core, it begins to utilize the helium it creates as a power source. Briefly, after hydrogen burning peters out, gravity once again gains the upper hand, crushing the core of the star, causing its pressure and temperature to rise. Once conditions are hot enough, the star begins to fuse helium nuclei together, creating heavier atomic species, such as carbon and oxygen. This newfound energy source causes the outer layers of the star to expand and cool. Thus begins the inexorable march towards stellar old age. The expansion of the star's outer atmosphere renders it far more luminous than it was during its main sequence era, sometimes as much as several thousand times brighter.

When you turn your telescope to **Aldebaran**, 65 light years away in the constellation of Taurus, you'll be eyeballing a star even more highly evolved than Capella. Its K5 spectral class is all too apparent, the telescope revealing its bright orange hue. Contrast that to **61 Cygni A**, one of a pair of showcase orange dwarf stars (both middle K spectral class) in the constellation of Cygnus the celestial swan. A small telescope at low power easily frames the duo, and though they shine with a similar hue to Aldebaran, there the similarity ends. 61 Cygni A lies squarely on the main sequence. It is orange not because it has bloated into stellar senescence but because it has a lower mass and hence a cooler surface. These binary stars will both outlive our own Sun by many billions of years.

It's worth taking an extra minute observing these aureal suns, for this was the first system to have its parallax measured way back in 1838 by the German astronomer Friedrich Wilhelm Bessel (1784–1846). Through painstaking work, often in lonely and freezing conditions, he achieved immortality by accurately measuring their distance. In so doing, Bessel provided us with the first true glimpse of the sheer vastness of interstellar space. The stars are frighteningly far apart!

Similar stories can be told when we examine stars of the M spectral class. First up is **Eta Persei**, a fairly inconspicuous star to the naked eye. Shining feebly overhead at magnitude +3.8, it is readily picked up owing to its deep orange coloration. This is an M3 star—an orange supergiant—some 1,300 light years away. But wait! If you take a closer look, you'll be able to make out a very faint spark of light, a companion of the ninth magnitude, tucked away with it against a picturesque backdrop of field stars.

By mid-winter, mighty Orion begins to assert himself, with the unmistakable red supergiant **Betelgeuse** (spectral class M2) dominating the eastern skyline. This is one of the true super-heavy weights of the stellar pantheon, a geriatric that will not end its days like more modest, sun-like stars. Betelgeuse will expire in a cataclysmic fireworks display. Its great mass (at least 20 times solar) will seal its fate.

Such stars burn helium into carbon and oxygen, as lower mass stars do, but the former then can undergo further rounds of nuclear fusion, furnishing still heavier atomic nuclei, such as magnesium, sulfur and silicon. But successive rounds of nucleosynthesis become less efficient for the star, and when the core fills with silicon, it sounds the death knell for these beautiful, giant suns. Silicon gets converted into iron, and iron does not burn (in a nuclear sense) at all. Because the synthesis of atoms with masses greater than iron requires more energy than they release by undergoing fusion, a fatal energy crisis ensues; the core of stars such as Betelgeuse will collapse under gravity and implode, becoming supernovae, shining briefly with the luminosity of a billion stars. Only the core of the star will remain after its violent death. We needn't worry, though; at over 600 light years away, we're at a safe distance to survive the pandemonium.

The electrons and protons making up the matter in the supernova remnant will combine to create neutrons, releasing a torrent of nearly mass-less particles called *neutrinos* in the process. The remnant of Betelgeuse will be like a giant atomic nucleus—a *neutron star*. And, just as a pirouetting skater spins more rapidly as she draws her arms inward, the tiny stellar core of Betelgeuse—only a few tens of kilometers across—will speed up its rate of rotation. While stars like the Sun revolve on their axis in time frames measured in days and weeks, neutron stars can complete their axial rotation in fractions of a second. The radio pulses emanating from the rapidly rotating neutron star, called a *pulsar,* sometimes happens to be directed along our line of sight.

Whether Betelgeuse will remain a neutron star or not is uncertain, as astronomers are divided about the precise mass of the core it will leave behind. You see, there is a limit to the amount of matter a neutron star can possess before gravity gets the upper hand. If the burnt out core of the supernova exceeds about two solar masses, then nothing can stop gravity from crushing it out of existence. When, or if that happens, Betelgeuse might become a black hole—a place where a great star once shone. It's mind boggling to think that this beautiful orange sun will 1 day be gone forever.

On the trail of decidedly different M class stars, we turn our attention to almost the opposite end of the sky. Sweeping the field of view with a low power (32-mm) ocular, you might come upon a faint ochre star of the seventh magnitude, on the border between Ursa Major and Leo Minor. Inserting a 9-mm orthoscopic yielding 84×, and centering, you should arrive at **Lalande 21185**, a faint, M2 dwarf star, yet still one of the brightest of its kind in all the heavens. These higher magnifications are needed to get a good look at its color, especially if there is a near full Moon risen in the east.

The fate of this star couldn't be more different to big, bold, Betelgeuse. With less than half the mass of our Sun, it will continue to shine, albeit, feebly, far longer than our own star. And though it seems rather inconspicuous through a telescope, it is one of the nearest stars to the Solar System, only 8.3 light years away, in fact. Such dim stars likely extend the plurality of worlds beyond our ken, for it is known that such systems can stably harbor planets and giant moons, some of which may be habitable. But life would be no bed of roses around such systems.

For one thing, any habitable planet around such a red dwarf star would likely be located very close to it, if it is to stably maintain liquid water on its surface. Indeed, it might often be so close as to be tidally locked, i.e., with one face of the planet or moon permanently facing the star and the other in total darkness—much like our own satellite, the Moon. Things would be very interesting along the day-night terminator of such worlds, where conditions might be a little more comfortable for such putative organisms. That said, these dwarf stars are far from quiescent. Indeed they are well known to emit dangerous X-ray flares, which could stymie the development of any advanced forms of life. And while they remain fascinating objects of study, we just don't know enough about them to say what they could really be like. But that's half the fun, isn't it?

Now let's look at an absolute corker. Enter **Mu Cephei**.

Known more famously as Herschel's **Garnet Star**, it's easy to find at the extreme western part of the constellation of Cepheus. This class M2 red supergiant star is a stunner in a telescope at 84×. It's fairly bright but can vary between +3.4 and 5.1, i.e., it's a semi-regular variable. It appears to be considerably redder than any of the other M stars nearby, a consequence perhaps of the prodigious levels of carbon-rich materials in its bloated, outer atmosphere. Such organic molecules are extremely effective at absorbing shorter visible wavelengths and allowing through longer wave (redder) light.

So there it is—life and death thrust haphazardly among the colorful stars. Not a bad way to spend a telescopic hour.

The Blessings of a Clear Sky

We are fortunate enough to live on a planet that yields clear skies. What might it be like for a sentient creature to live on a planet permanently shrouded in cloud? These folk would never see the stars. They could build tools and cities. They would have to figure out how to tell the time and recognize the passage of the seasons by some other means not involving the position of their sun or the stars in the sky. They would never develop a natural curiosity for the stars, because their culture could never absorb the marvels of the heavens.

Such creatures would sooner or later discover the atom, Newtonian physics (by some other name of course!) and the principles of quantum mechanics. They might eventually figure out that *something* lies beyond the clouds, but without knowledge of a vast and ancient universe—information that could be readily gleaned from just observing the stars—this extraterrestrial society would be severely set back in their motivations for exploring the cosmos.

Think of how different we might have been had we lived on a world where the daytime glory of the Sun and the lesser splendor of the stars in the night sky were never visible. These visual wonders have been at the center of our cultural and spiritual inclinations for hundreds of millennia. Our ancestors patiently watched and recorded the motions of the Sun, Moon and distant stars, with the passing of the seasons. Much of the religion and myth of our species is centered on the celestial bodies. The Sun and the stars have profoundly influenced our mathematical discoveries, their rising and setting painting the impression of a great circle in the sky—a celestial sphere. The nearly circular faces of the Sun and full Moon must have inspired some mute and forgotten mathematician in prehistory to contemplate the geometry of the circle and the other conic sections.

Many of our concepts of the nature of light and other forms of electromagnetic radiation have been inspired by studying the stars. Even some elements—helium for example—were discovered not on Earth but in the Sun. Indeed, other species of Planet Earth have developed a deep affiliation with the stars. Some birds, such as migratory geese, for example, navigate at night by their light. They can identify the positions of many of the brighter luminaries in the sky and can apparently compensate mentally for their changing positions as the night progresses.

The stars above have shaped our evolution from hunter-gatherers to farmers, and from technologists to spacefarers.

They are as much a part of us as we are of them.

Thank goodness for a clear sky!

Diligent research conducted over the last century has given us a fairly good model of how stars come to be. They are hatched from interstellar gas—mostly hydrogen—and dust that, over time, becomes progressively more concentrated. During the early stages of star formation, the attractive force of gravity molds an amorphous cloud into a spherical shape. As the cloud collapses under its own weight, it releases energy that heats the gas and dust until it starts to glow, weakly at first, in the infrared. And as it contracts and heats up still more, it begins to emit visible light, and a protostar is formed. As the interior of the collapsing sphere grows hotter and denser, the bonds holding the molecules of hydrogen together are broken, and then the individual atoms lose their electrons to become electrically charged particles called *ions*. This hot soup of negative and positively charged particles is called a *plasma*. In the final stages, when the pressure is high enough and temperatures exceed 15 million K, the individual hydrogen nuclei slam into each other, creating helium.

Fig. 7.4 Majestic Orion, with the Great Nebula (M42) blazing below the central star in the Hunter's Belt, also called the Sword Handle (Image by the author)

Nuclear fusion liberates enormous amounts of energy. If you were to carefully weigh a helium nucleus and compare it to that of the hydrogen nuclei out of which it is forged, you'll find that there's a small mass deficit, the former being 0.7 % lighter than the combination of the latter. This lost mass is converted into energy, lots of it. Indeed fusing 1 kg of hydrogen into helium releases the same amount of heat as burning 20,000 tons of coal!

The youngest and most massive stars release a torrent of high energy ultraviolet light from their surfaces, so much so that they cause the gas in the surrounding cloud to fluoresce. This is what astronomers call an *emission nebula.*

Though astronomers have identified many star-forming regions peppered along the spiral arms of our galaxy, the nearest one to Earth that spawns massive stars is the Great Nebula in Orion (M42), located on our cosmic doorstep, as it were, just 1,400 light years away.

The story of M42 is one of enduring mystery and fascination. Easily visible to the naked eye—even from a light-polluted location—it can be seen as a fuzzy, fourth magnitude patch immediately beneath the middle star in Orion's Belt. Intriguingly, though he compiled an accurate star atlas that included the fainter Andromeda Galaxy, the tenth-century Arabic astronomer Al Suffi failed to record it. Neither was it mentioned by Johann Bayer (1572–1625) in his 1603 magnum opus, the *Uranometria.*

Weirder still, Galileo Galilei, sweeping this region of sky with his telescope in December 1610, only noted an increase in star density in the locus we now know is occupied by M42. And yet, more recent research suggests that the native peoples of Meso-America gave mention to M42 as 'smoke from a campfire' in sources that predate the arrival of Columbus in the New World.

Just why this chronicle of events should be is unclear. Was the nebula clearly visible to the pre-Colombian Central Americans? Was the nebula invisible through-out the pre-telescopic era? Did it, quite suddenly, 'appear,' as if out of nowhere, sometime in the seventeenth century?

Alas, nobody knows for sure, but what we do know is that less than a year after Galileo's famous 'non-discovery' of the Orion Nebula, the Frenchman Nicholas Peiresc (1580–1637), described it as a 'small illuminated cloud.' And in 1618, the Jesuit astronomer, Johann Baptist Cysat (1587–1657) likened it to a comet that was seen close by about the same time. It was not until 1656 that the Dutch scientist Christiaan Huygens (1629–1695), using a Galilean telescope, provided the first detailed description of the nebula. His famous sketch, outlining the brightest part of the nebula (actually about 3 light years across), is still referred to as the Huygens' region today.

Even a 60-mm refractor, at modest enlargements, will easily show you a fetching quartet of stars at the heart of the nebula—the so-called *Trapezium*—yet Huygens only recorded three! And his telescope was more than capable of resolving the fourth member. Could this be further evidence that this magnificent nebula was gradually set alight in historical times?

Sir William Herschel excitedly turned his first successfully made speculum reflector on M42 in the winter of 1774. And though the great observer thought all nebulae could be resolved into mounds of stars, even his 40-foot reflector failed to do so. It was Sir William Huggins, of spectroscopy fame, though that put Herschel's theory to rest, when he showed that the emission spectrum of the nebula was quite unlike that of any star. And while it was clearly at the top of the observing list for many astronomers throughout the centuries, Admiral Smyth (1788–1865), observing in the nineteenth century, expressed a rather ambivalent opinion concerning the nebula, describing it only in passing as an object encasing the magnificent Theta Orionis complex. Smyth was however, a double star observer after all!

A little grab 'n' go telescope of 80-mm aperture is more than capable of gleaning surprising detail from this wonder of the celestial realm. For this study, just two eyepieces are to be employed—a 40-mm Erfle yielding a 3.5° true field at a magnification of 19×, and a 9-mm hyper-wide angle (100° apparent field) ocular, delivering an enlargement of 83× in a 1.25° true field.

It is most difficult to describe the feelings that come over you, having just centered the great nebula in the field of view. Suffice it to say that your heart rate will increase just a little, as you bring the object into sharp focus. The English amateur astronomer, Reverend Webb, observing it in the nineteenth century, described it as 'an irregular branching mass of greenish haze.' That may not be completely inaccurate. In the 19× enlargement of a small telescope, the object appears pale green, appearing as a dove with outstretched wings, orientated northwest through southeast and occupying an area of sky nearly twice the size of the full Moon. At its heart (embodied by the famous Trapezium), and sprinkled across its shining countenance, lie numerous young, hot stars that render it visible.

This author has coined the word 'chlorescent' to describe its unique color, as it is the recombination energy of oxygen ions and their electrons that gives rise to the pale green light ('chloros' means green in Greek) emanating from the nebula on the darkest, most transparent nights. This is a most anthropocentric perspective, however, as electronic eyes, with their greater sensitivity to longer visible wavelengths, invariably present the nebula in a riotous splash of pink and red. Immediately to the north of the main structure (near the so-called 'fishes mouth'), a careful glance will show you **M43** (De Mairan's Nebula), supposedly a separate entity but actually an integral part of a much larger complex known to professional astronomers as the Extended Orion Nebula (EON).

With averted vision on the darkest nights, you can trace the luminous filaments quite a bit away from the main structure, although it pales into insignificance compared to a higher power, wide angle view. A 9-mm 100° field eyepiece, coupled to the SV80/9D, will provide a killer view for a 3-in. glass; it's the magical combination of hyper wide field and high magnification acting synergistically to help a curious eye see well.

The cosmos is well past its heyday for fashioning stars. That happened 11 billion years ago. Indeed it's sobering to think that over 90 % of all the stars that will be born have already been born. Centering the Trapezium in the field of view and carefully focusing, you can make out four white stars, labeled A through D. Spectral analysis reveals that they are very young and hot and so cannot be more than a few million years old (and Star C is probably significantly younger still). Radiation pressure from the hotter components of the Trapezium (mainly Theta C) has pushed away some of the gas and dust, attenuating the nebula in its immediate surroundings and rendering a little cavity, a tiny blister, with its sweet surprise, visible to us.

Of the four bright stars your telescope can see, three of them are binary—very close binaries and hopelessly beyond the reach of a garden spyglass. C is solitary. But have a good, long look at it. It is probably only a few thousand years old and has the honor of having the hottest stellar surface temperature yet measured. One sometimes wonders whether Huygens missed this star (you'll remember he only saw three of the four Trapezium stars) because it was not yet visible to him? Star B is an eclipsing binary system, i.e., the companion brightens and dims the other member periodically, as it laps it. This cycle of events happens in under a week (6.5 days to be more precise), causing the primary to fade by 0.5 stellar magnitudes. Star A follows suit, though it takes ten times longer for it to complete one cycle, its light being diminished by as much as one stellar magnitude over a 20-h spell. How would one know it's in eclipse? Well, if it appears dimmer than star D (normally its equal), then you've caught sight of it. This event has been noted on and off a few times over the years.

Astronomers know of four or five other stars buried within the Trapezium complex, most notable of which are the E and F components. Both of the tenth magnitude, they are difficult to spot in a small telescope. But not so much for the lack of aperture—it has enough to bag such faint objects from a dark sky—but because of the proximity of both the E and F stars to the much brighter A and C components, respectively. Good seeing and high magnification are the tools of the trade.

The telescope paints a tidy little picture, as if four eggs were neatly laid out in a basket fashioned from wafer-thin rice paper. But that's just a human convenience. Those four stars are enormous by solar standards (class O and B), each some 15–30 solar masses and occupying a volume of space only 1.5 light years from edge to edge—stars that are doomed to go boom in the not too distant future. And, if that weren't perilous enough, there's likely a 150 solar mass black hole brooding away at the cluster's center, according to a discovery made in November 2012. You can't see it, but you *can* observe (spectroscopically) the peculiar velocities of its attendant stars as they course through Orionid space.

Much finer details of the nebula itself can be discerned with the high power Oculus. One can trace the nebulous filaments nearly as far as NGC 1977, located half a degree to the north of M42. Away it meanders southwards, too, almost coalescing with the glorious triple star **Iota Orionis**. NGC 1977, otherwise known as the **Running Man Nebula**, appears as a wreathlike conglomeration of faint suns, with the brightest luminaries, 42—and 45 Orionis, regally gleaming at the fifth magnitude.

On the finest nights, with full dark-eye adaptation and with averted vision, one can just make out some tufts of nebulosity round these brighter members. Other observers, enjoying views of the nebula from more advantageous southerly latitudes, will have no trouble seeing more than average. And the same is true of the nebulosity skirting with Iota Orionis (**NGC 1980**) to the south.

To gain any further understanding of this extraordinary place, we need to talk to the animals. Vipers have fangs that can track down the heat from their warm-blooded prey and Cetacean mammals emit high frequency sound waves (ultra-sound) to construct a three-dimensional picture in their heads of their murky surroundings, as well as creating stun guns for hunting.

Animals do what they do because they can. Humans were endowed with technological intelligence to do what other species can do and more. We can design detectors that can observe astronomical bodies at other, normally invisible wavelengths, to glean more information. And it's not guesswork. Stars are simple things; their aspects can be modeled mathematically. And, as we've previously seen, there is a relationship between the range of colors emitted by a luminous body and its temperature. This is particularly true when the object is in thermal equilibrium, absorbing and emitting radiation equally well. Such an idealized object is referred to as a *blackbody*.

Stars are blackbodies, or very nearly so. And it turns out that there's a very simple connection between the wavelength at which a blackbody shines most brightly and its temperature. The cooler the temperature the longer its peak wavelength will be (Wien's law). That's why red stars are cooler than blue stars. If star formation is occurring, we could use longer wavelengths to witness the nativity of young suns not yet set alight by nuclear fusion. That's where infrared and sub-millimeter astronomy can play a powerful role in seeing deep into mysterious places, places just like the Great Nebula in Orion.

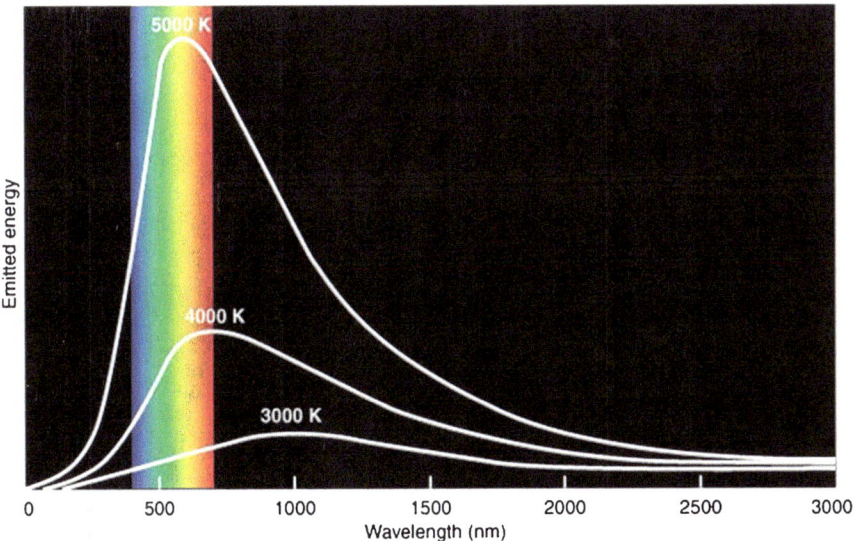

Fig. 7.5 Wien's law, which shows how peak radiation output is inversely related to absolute temperature (Wiens law of radiation, http://en.wikipedia.org/wiki/File:Wiens_law.svg)

Placed high above the atmosphere, where these radiations are not absorbed by water vapor, infrared telescopes have transformed our knowledge of this and other star-forming regions. Penetrating the obscuring dust, telescopes such as the Spitzer Space Telescope and the Herschel Space Observatory have revealed this place to be utterly violent and unruly. Countless thousands of protostars, in a kind of cannibalism, are eating the Extended Orion Nebula (EON) inside out, consuming its vast reserves of gas and transforming it into stars. The most massive of these produce violent winds, directed by strong magnetic fields, as well as murderous radiation that would sunder a living thing into smithereens faster than you could say 'pandemonium.'

Star formation is far from pretty, but we have to take the rough with the smooth.

Speaking of smooth, isn't it marvelous that we live on a wholesome world largely buffeted from what seems to be routine cosmic violence. Thank goodness for small mercies!

The Great Stellar Diaspora

The laws of physics are the same today as they were yesterday, and they will be the same tomorrow. They allow us to make sense of the world around us, to elucidate the chronicle of events that led from the violent birth of stars to the establishment of star clusters and their attendant planetary systems. Our journey continues with an exploration of a few modest clusters easily accessible to the busy grab 'n' go astronomer.

No star is born in isolation. As we have seen in our study of M42, the Great Nebula in Orion, they are conceived inside enormous interstellar clouds of gas and dust, which, depending on their size, can fragment and give rise to anything from a few dozen stars to many thousands. When the behemoths of the stellar world—the brazen O and B stars—have died away, the Orion Nebula will 1 day become a stable cluster, brimming with stars of the A, F and G spectral classes. Massive stars, owing to their short life spans, live out their entire lives at or near the places of their births. Smaller suns, enjoying far greater longevity, have ample time to migrate from their place of birth. They are the ultimate cosmic migrants. What's more, the larger the molecular cloud, the more stars that are born and the greater their gravitational influence over each other. Thus, larger stellar nurseries take longer to disperse than their smaller counterparts.

Once the last of the gas and dust dissipates, an *open* or *galactic cluster* is born. On a dark, moonless night, you may be forgiven for thinking that they—over a thousand strong—are strewn randomly across the sky. But detailed surveys over the years reveal that the vast majority lie within a thin, angular strip, some 25° above and below the galactic equator. Some parts of the sky seem richer than others owing only to the paucity of light obscuring interstellar dust in those directions. The perspectives of Cassiopeia and Puppis are particularly noteworthy in this regard, veritable open cluster oases in the starry firmament.

Fig. 7.6 The glorious Double Cluster in Perseus (Image by the author)

As time further marches on, the cluster becomes influenced by tidal forces from others nearby, as well as shock waves that ripple through the tenuous matter between the stars (the so-called interstellar medium). This results in some of the smaller and less massive members of the cluster, picking up speed, eventually breaking loose from the center of gravity of the system.

Why the Diaspora of stars from their place of birth? To answer that, we must first take a closer look at the great diversity of open clusters that decorate the sky.

First up, the famous Double Cluster in Perseus, located high overhead at this time of year and easily located as a fourth magnitude fog, about halfway between Delta Cassiopeiae and Gamma Persei. Famously unlisted in Charles Messier's celebrated catalog, it was however placed (as entry number 14) on the late, great Sir Patrick Moore's list of Caldwell objects.

Though our eyes can make little out, save a luminous smudge between the great northern constellations, the 3-in. glass is an inordinate help, for even in the expansive, low-power field of an Erfle eyepiece, both clusters—NGC 884 and the more easterly NGC 869—take the eye by storm. Located about 7,300 light years away—nearly five times farther away than the EON—its brilliance dazzles the dark-adapted senses, even though one is only seeing a smattering of a few dozen stars in each cluster. And while some astronomers argue about a possible age difference between these clusters, they are, more than likely, cut from the same cloth; and only 4–5 million years old!

Could this be anything like the view our distant descendants might see when they cast their gaze towards Orion's Belt? This already magnificent Orion Nebula would be rendered more glorious still, owing to its much greater proximity to us. If so, amateur astronomy is going to get a lot more interesting!

Both clusters in Perseus are about the size of the full Moon and are separated by a 0.25° dark interface. With the low power eyepiece, it is exceedingly difficult to clearly make out any real color differences between the stars in these clusters, but when the 9-mm Oculus is inserted, the drought ends. In that relatively enormous 83× field, many more stars can be seen, and with that, a few distinct stellar hues become apparent—notably steely white, yellow and red. The westernmost cluster, NGC 884, in particular, presents a splash of red stars, like tiny poppies in a distant field. This cluster, only a few million years older than the Orion Nebula, is home to red giant stars, stars that are already heading for their graves! What a difference a couple of million years makes!

Might this suggest that NGC 884 is the more evolved of the two? Perhaps, but then again star formation can occur at different times within the same cluster, so we don't know. Remarkably half the members, which number between 150 and 200 for both clusters, are of a special type of variable star (a subject we will examine in far more detail later) stars that are characteristically young and therefore indicative of the salubrity of the clusters. In fact, they have barely rid themselves of the thick shrouds of gas and dust that had rendered them all but invisible just a few thousand millennia back.

C14 lies in a different spiral arm to that on which our Sun and the EON reside (the Orion Arm) and, as such, its glory is diminished somewhat by intervening dust that litters the interstellar medium between them. By carefully observing the cluster at higher magnifications, you'll be able to see many dark, cavernous places winding through both clusters, where the stars don't shine. Were it not for that dust, the Perseus Double Cluster would be on the tip of everyone's lips—like **M45**, the immortalized **Pleiades**, the object to which we turn to next.

Instantly recognizable to even a casual sky gazer, the Pleiades, or Seven Sisters, is arguably the most famous, brightest and glorious open cluster in all the heavens. It is easy to find as a miniature version of the Big Dipper in the winter sky.

Known to the ancients, even a casual glance from a fairly light-polluted sky shows six components, though a more concentrated gaze from the same location usually returns seven. From a very dark sky, as many as 10 stars can be made out, and there are some people who can count 13 members and upwards. How many can you see with your unaided eyes? Bizarrely, Messier included the Pleiades in his list, even though everyone knew it was not a comet.

Binoculars or a rich field telescope transforms this naked-eye delight into a blaze of stellar glory, with several dozen white and blue–white stars being easily resolved, even at low power. Long exposure images of the Pleiades reveal an intense blue wisp around many of its brightest stars. At first it was thought that this must be the last vestiges of the gas and dust out of which the cluster formed, but we now know that the whole cluster is embedded in a huge dust cloud that reflects the light from the hot blue–white stars, i.e., it's a *reflection nebula*.

Fig. 7.7 The majestic Pleiades cluster (M45) (Image © Mike Pearson. Used with permission)

On the darkest, most transparent nights, close scrutiny of Merope, one of the cluster's main stars, reveals this reflection nebula rather well in large binoculars or a small telescope. It's well worth the effort to sit patiently on a good night to see if your eye can spot it. In 1859, the German amateur astronomer Wilhelm Tempel (1821–1889), who clearly saw it with a 4-in. achromatic refractor, famously described the view as like that of a 'breath on a mirror.' He wasn't far wrong!

The 40-mm Erfle presents a field positively chock full of stars, radiating in glory from the 3rd visual magnitude right down to the ninth or tenth. Unlike the clusters in C14, you'll see that the cluster is well opened up, even at its relatively close distance of about 400 light years. Plotting its stars on an HR diagram reveals the cluster to be about 50 million years old, with virtually all the bright stars (at least 100) squarely on the main sequence. That extra time has enabled this galactic cluster to unravel itself, as well worked-out theory suggests.

When the central portion of the Pleiades is examined with the Oculus, the eye is assaulted with an avalanche of lux. And even though the eyepiece cannot conceivably take in the entire cluster, it is quite an experience just to let those brazen Pleiades members drift through the field. What is most striking is that the Oculus doesn't show many double stars—especially close ones—in this cluster.

Fig. 7.8 The evolved Beehive Cluster (M44) in Cancer (Image © Mike Pearson. Used with permission)

As much as one can see with a small telescope, there is much about the Pleiades that the eye cannot discern. For example, some of its most prominent stars are rapidly rotating, some up to 150–300 km/s. That comes with the territory for hot B stars. Utterly blind we are, too, to the swarms of tiny stars inside the cluster, stars so small that they do not shine in visible light but only in the infrared.

In 1995, astronomers discovered the first so-called brown dwarf inside the Pleiades. Belonging to a class of stars called the L dwarfs, these diminutive bodies typically only contain about 6 or 7 % of the Sun's mass (so 60–70 Jupiter masses), yet are probably smaller than Jupiter, owing to their high density. Though exceedingly numerous in the cosmos, their cores never quite achieve the temperatures and pressures required to begin hydrogen fusion (though deuterium—an exotic form of hydrogen—fusion is thought to occur in the smaller brown dwarfs, as well as lithium fusion in the largest examples). Either way, they'll probably last forever, or nearly so!

Turning now to the stars of **Praesepe**, the celestial Manger (**M44**), are a much more ancient and evolved open cluster, and the jewel of the celestial Crab.

Shining with an integrated magnitude of 3.1, it is easily spotted about midway between the bright stars of Gemini (Castor and Pollux) and Regulus in Leo. Galileo elucidated its true nature in 1610 when he recorded some 40 members of the cluster. A 3-in. glass can show twice as many brighter than magnitude +10. Its most luminous members shine modestly at the fifth and sixth magnitudes and provide a very fetching view in a 3.5° field of the author's 40-mm Erfle. The cluster—which is loosely bound at a glance—sprawls haphazardly over 1.5° of sky.

The Oculus does a superb job of pulling many of the fainter members out of the background murk. In contrast to the Pleiades, Praesepe is chock full of double stars, with many of them worthy of scrutiny with a larger telescope. You'll also notice that this cluster, estimated to be located about 500 light years away, has a number of prominent orange giant stars that have evolved off the main sequence. Indeed, HR diagrams of M44 indicate that it has an estimated age of about 500 million years.

There are many other clusters in the sky considerably older than Praesepe, but these three—C14, M45 and M44—serve a useful purpose in teaching us how they evolve over time. Our Sun was conceived in a clutch of stars some 10,000 strong about 4.6 billion years ago. It underwent the same tumultuous scheme of events from cloud fragmentation, protostar formation and migration away from its birth cluster. After nearly 5,000 million years, the Sun's siblings, stars hatched from the same giant molecular cloud complex, are hopelessly dispersed throughout the galaxy. But it would be dead wrong to think of the Sun as an ordinary star. New and exciting research over the last two decades has cast severe doubts over Carl Sagan's now almost clichéd vision of a 'humdrum star,' lost somewhere between 'eternity and immensity.'

For now, though, it's time to pause, take stock, suggest some other activities and pose questions to help you consolidate some of the ideas we have covered thus far.

Activities

Colors of Stars

Move your telescope to a random star in the sky. Examine it at low, medium and high power. What color is it? If you have difficulty seeing any color, try defocusing the star ever so slightly. Does it have any companions? If so, are they close by or wide away? Carry out further observations on another 14 or so, stars. Where they are in the heavens doesn't matter.

The best views are to be had up high in the sky, where the air through which you observe is thinner. You can do this many, many times. Each night, 15 new stars. Make notes of what you find.

How many red stars did you come across? How many orange, yellow, white and blue–white stars did you see? How many of them had companions?

Three More Clusters

If you're lucky enough to see the constellation of Canis Major high in your sky, then you'll do well to track down **Caldwell 64** (NCC 2362), about 13° southeast of bedazzling Sirius. In the right southern location, a small telescope will reveal a monster star, **Tau Canis Majoris**, embedded in a tight cluster of fainter members, some 60 strong. Though it shines modestly at magnitude +4, it's awfully far away—over 5,000 light years in fact. Use moderate powers to resolve the components. Tau Canis Majoris is a blue supergiant (one of several O and B stars in the cluster) and so dates the cluster to just a couple of million years. Indeed, it could just turn out to be the youngest of its kind in all the Galaxy.

Exploring the Coma Berenices Star Cluster

East of Leo's hindquarters and west of the bright star Arcturus lies an often over-looked and rarely celebrated star cluster, the outline of which is visible to the naked eye. Look for a foggy concentration of stars, roughly wedge-shaped pointing northward and about 5° in angular extent. What you're gazing at are the brightest members of one of the closest open clusters in the heavens—the **Coma Berenices star cluster** (known also as **Berenice's Hair**, or Melotte 111).

Known since antiquity, the true nature of this grouping of stars remained uncertain for a long time. Philibert Jacques Melotte included this object in 1915 in his *Catalog of Star Clusters* as Melotte 111. Finally, in 1938, Robert Julius Trumpler proved that Melotte 111 was a true cluster by demonstrating that at least 37 members have a common space velocity. The latest calculations have put the cluster at a distance of 288 light years. Indeed, only the Hyades and the Ursa Major Cluster are closer.

With even the most humble of optical accoutrements, the view is simply marvelous. The cluster takes up most of the field of view in a 10×50 binoculars and is all the more striking because sky background looks devoid of faint field-stars. That's not surprising when one learns that Melotte 111 lies only a few degrees from the north galactic pole. The cluster can be observed by a wide variety of instruments, but an exceptional was enjoyed by this author, using an ultra-Portable 3-in. (76-mm) F/6 refractor and a power of 15× that presented a wonderful 5.3° field of view. With such an instrument, the entire asterism drizzles the entire field in starlight, both delicately faint and gloriously bright. The lovely high-contrast view shows about 150 stars, but looks can be deceptive. Only the brightest stars comprise the true moving cluster; the many fainter suns being mere background field stars. Thus, Melotte 111 benefits greatly from chance alignments of unrelated objects. Many of the stars in the cluster seem detached, singular, almost lonely. But there are a few notable exceptions, 17 and 16 Comae, in particular. Cranking the magnification up to 34×, giving a 2° field in the little refractor, 16 Comae shows itself as the central jewel surrounded by a quartet of fainter suns, each about eighth or ninth magnitude in brightness.

In terms of age, Melotte 111 is a young cluster, roughly 400 million years old. That's evident from the paucity of older, red stars in the asterism. Instead, we invariably find early spectral-type main sequence stars (B–F). The very loose impression the cluster makes on inspection reflects the fact that the cluster is rapidly dispersing into the Galactic disc and may all but disappear in a few tens of millions of years from now.

Digging Deeper

It is often said that Melotte 111 is a low power asterism, an object worthy of study with small, rich field instruments. But that's not entirely so. With a 6-in. reflector, about a dozen background galaxies are visible within a field of 2° centered on the main concentration of starlight, including two Caldwell objects (**C36** and **C38**). All shine dimly but distinctly at magnitude 10–12.

Of course, if you survey the region with a large light bucket such as a 12-in. (305-mm) F/5 Dobsonian reflector, endless new treasures—mostly of a galactic nature—come out of the darkness to greet you. Indeed, the perceived emptiness of the region surrounding the cluster is transformed into a rich hunting ground for galaxies of all types. That much is made clear if you pan southward towards the Virgo bowl, where literally hundreds of faint fuzzies puncture the darkness with their light. Indeed, these island universes—some 3,000 in number—make up the famous Coma Berenices galaxy cluster, strewn across 20 million light years of empty space.

So, what are you waiting for? Whether you use your eyes, binoculars, a small rich field instrument or a giant light bucket, spend some time caressing Berenice's Hair. She wouldn't mind your gentle approaches.

Investigating Stellar Old Age and Death

It is not an inconsiderable fact that our Sun, now in its middle age, is now at its most stable, allowing human civilization to thrive. But it will not remain so. Slowly, over millions of years of time, our star will evolve into old age and finally die. The same is true for all the other stars that decorate the sky. In this section, we'll be exploring the subject of stellar old age and death and take a look at some celestial targets that will affirm the concept of stellar evolution.

As the hydrogen around the core is consumed, the core absorbs the resulting helium, causing it to contract further, which in turn causes the remaining hydrogen to fuse even faster. This eventually leads to initiation of helium fusion in the core. In stars of more than approximately solar mass, it can take a billion years or more for the core to reach helium ignition temperatures.

Fig. 7.9 Stellar evolution of a sun-like star from the main sequence (*left*), through the red giant phase (*middle*) and finally ending its life as a planetary nebula (*far right*) (Image courtesy of ESO)

When the temperature and pressure in the core become sufficient to ignite helium fusion, a *helium flash* will occur if the core (for stars under 1.4 solar mass) is largely supported by *electron degeneracy*—a 'pressure' caused by the requirement of electrons to obey the rules of quantum mechanics. In more massive stars, the ignition of helium fusion occurs relatively quietly. Even if a helium flash does occur, the time of very rapid energy release (on the order of 10^8 suns) is brief, so that the visible outer layers of the star are relatively undisturbed. The energy released by helium fusion causes the core to expand, so that hydrogen fusion in the overlying layers slows and total energy generation decreases. The star contracts and migrates to the horizontal branch on the Hertzsprung–Russell diagram, gradually shrinking in radius and increasing its surface temperature. Core helium-flash stars evolve to the red end of the horizontal branch but do not migrate to higher temperatures before they gain a degenerate carbon-oxygen core, which, like the degenerate electron rich cores of lower mass stars, allow them to initiate helium shell burning. These stars are often observed as a red clump of stars in the color magnitude diagram of a cluster, hotter and less luminous than the red giants. Higher-mass stars with larger helium cores move along the horizontal branch to higher temperatures, some becoming unstable pulsating stars in the yellow instability strip (RR Lyrae variables), whereas some become even hotter and can form a blue tail or blue hook to the horizontal branch. The exact morphology of the horizontal branch depends on parameters such as the star's chemical composition (what astronomers refer to as its *metallicity*), age, and helium content, but the exact details are still being modeled.

After a star has consumed the helium in its core, fusion continues in a shell around a hot core of carbon and oxygen. The star follows the asymptotic giant branch on the Hertzsprung–Russell diagram, paralleling the original red giant evolution but with even faster energy generation (which lasts for a shorter time).

Although helium is being burned in a shell, the majority of the energy is produced by hydrogen burning in a 'higher up' shell, closer to the surface of the star. Helium from these hydrogen burning shells drops towards the center of the star, and periodically the energy output from the helium shell increases dramatically. This is known as a *thermal pulse*, which occurs towards the end of the asymptotic-giant-branch phase. Depending on mass and composition, there may be several to hundreds of thermal pulses, which lift the outer layers of the star, accelerating mass loss and transforming the star into a red giant.

Activity: Observing Some Red Giant Stars

Here is a list of common red giant stars. Use your telescope to examine them at low and medium power. Note their ruddy colors.

Star	Spectral class	Distance (light years)
Aldebaran	K5	65
Betelegeuse	M2	650
Antares	M1	550
Beta Andromedae	M0	200
Mu Geminorum	M3	230
Mu Ursa Majoris	M0	250
Delta Virginis	M3	200
R Lyrae	M5	350
R Cassiopeaie	M7	350
19 Piscium	C5	760
30 Herculis	M6	360
R Aquarii	M7	640
R Leporis	C7	820
Mira	M7	420
R Leonis	M8	325

Find out how many of the stars in the table above are variable. What might account for their variability?

Sun-like stars, having evolved past the red giant phase, have converted all their hydrogen and helium into carbon and oxygen. Further rounds of nuclear fusion are no longer possible as their core temperatures never rise high enough. The outer layers of the star begin to pulsate, but much of the material escapes the gravitational field of the star. This shell of ejected material forms the basis of a new *planetary nebula,* lit up by the fierce radiation emanating from the hot stellar core at its heart.

Many of these planetary nebulae possess multiple shells, due to the red giant experiencing several bouts of pulsations. Strong stellar winds, intense magnetic fields and the presence of disks of material girdling the equator of such geriatric stars are all thought to be responsible for explaining the wide variety of morphological forms these nebulae exhibit.

Telescopically, planetary nebulae tend be small, compact objects that, upon casual inspection, resemble planets. That's why the name was first coined by Sir William Herschel, who likened their appearance to the planet Uranus.

What remains after the outer layers of the star are shed to the surrounding interstellar medium is the burnt-out stellar corpse of very high density—a *white dwarf.*

A Famous Planetary Nebula in Lyra

When introducing youngsters to the wonders of the night sky, one of the first things you can turn to is the glorious **Ring Nebula** (**M57**) in the constellation of Lyra, the celestial Lyre. Favorably placed for observation for much of the year from the British Isles, this unusual object is easy to find. First locate the bright white star Vega and pan a few degrees east southeast until you arrive at a pair of 3rd magnitude stars known as Gamma and Beta Lyrae. You'll need a small telescope and moderate magnification to find the Ring Nebula, which is located about three-fifths of the way from Gamma to Beta. A 3-in. (76-mm) 'scope does a good job pulling it out of the background sky. Indeed, it was a similar sized instrument that the French comet hunter, Antoine Darquier, discovered the nebula in 1779, describing it as "very dull but perfectly outlined; as large as Jupiter and looking like a fading star."

Because of the object's small size (one arc minute) and relative faintness (magnitude +9.7), a 4-in. 'scope is about the minimum required to study this amazing structure in any detail. Though not a necessity, attaching either an OIII or UHC filter to the eyepiece also helps bring out further details. At 50× in a 4-in. (102-mm), you can easily make out the famous smoke ring appearance of M57. Like a doughnut, the central region appears hollow, almost cavernous, while the rim is brightly luminous. At high powers (150×) in the same instrument, the ring takes on a more elliptical structure, and you can begin to see that the luminous outer ring, which takes on a pale blue-green hue, is unevenly illuminated. With a really large telescope and high magnification, though, the non-homogeneity of the luminous gas that makes up M57 becomes obvious. For example, using a 12-in. (305-mm) reflector at 300×, the two sides of the ring resolve into curved and twisted filaments of gas.

Instruments with an aperture of 12 in. or greater are required for you to have any chance of seeing the central star—a burnt-out white dwarf at the end of its life. It can sometimes be spotted with averted vision with a 12-in., but the chances of success very much depend on the sky conditions. The slightest bit of haze or atmospheric turbulence renders it completely invisible, so be sure to wait for a good dark night to make your observations.

Once you've had your visual fill of the Ring Nebula, think of the journey we've made in unraveling its true nature. Sir William Herschel believed it to be a vast torus of stars that were just beyond the resolving power of his instruments. That said, the Third Earl of Rosse, observing M57 on many occasions throughout the 1840s and using the 72-in. Leviathan at Birr Castle, Ireland, couldn't resolve it into stars, either. Clues to its true nature had to wait until 1864, when the wealthy Londoner, William Huggins (1824–1910), examined the spectra of multiple nebulae, discovering that some of these objects, including M57, displayed the spectra of bright emission lines characteristic of fluorescing glowing gases. Huggins concluded, correctly as it turned out, that most planetary nebulae were not composed of unresolved stars, as had been previously suspected, but were something else entirely. The final piece of the puzzle had to wait until a proper theory of stellar evolution emerged in the early twentieth century.

The Ring Nebula, located some 2,000 light years away, is a vast bubble of spent nuclear fuel plowing through the cold dark of interstellar space after being shed from a geriatric star some 20,000 years ago. The slowly cooling *white dwarf* at its center emits a torrent of high-energy UV rays, lighting up the nebula from within and causing it to glow in the most gorgeous pastels of green, red and blue—if only your eye were sensitive enough to see them in their full glory! Knowing how far we've come with M57 makes observing this magnificent object even more worthwhile, so why not go out tonight and take a look for yourself?

Observing Other Planetary Nebulae in the Sky

The **Eskimo Nebula** (**Caldwell 39**) in Gemini is an excellent example of a planetary nebula that exhibits structure when examined at high powers. To find it, pan your telescope about 2.4° east-southeast of magnitude +3.5 Delta Geminorum. An 80-mm refractor picks up the Saturn-sized (15") oval shining at magnitude +9.1. Cranking up the power to 83× allows you to see a bluish shell of luminous gas surrounding a star-like object (magnitude ten) at its center.

Pause for thought. Though the Eskimo Nebula lies about 3,000 light years away, the central star, a young white dwarf, is only the size of Earth!

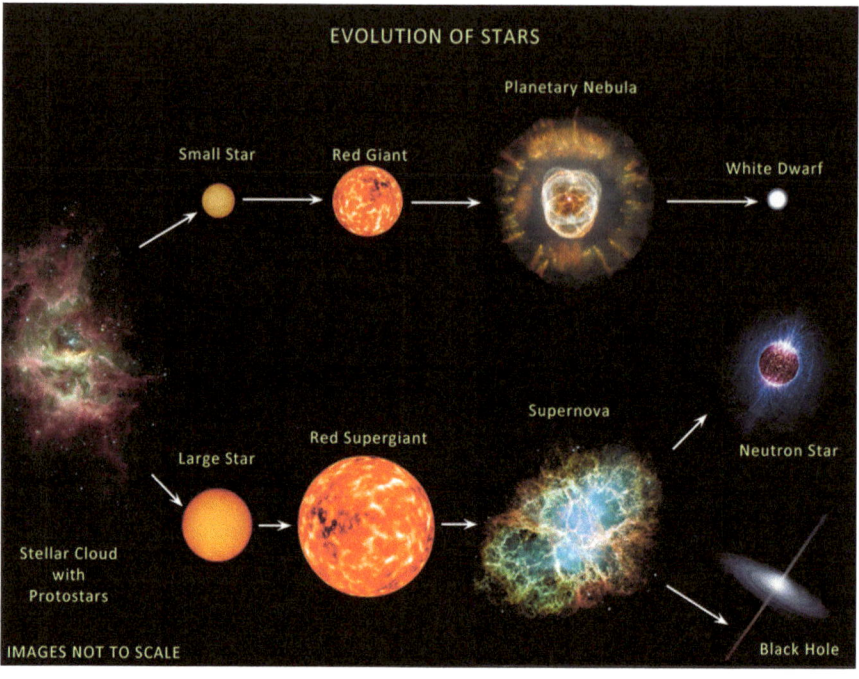

Fig. 7.10 Summary of stellar evolution (Image courtesy of NASA)

Here's a list of other prominent planetary nebulae that can be observed in a small, grab 'n' go telescope.

Name	Constellation	Magnitude
Helix Nebula	Aquarius	+7.6
Saturn Nebula	Aquarius	+8.0
Owl Nebula	Ursa Major	+11.0
Dumbbell Nebula	Vulpecula	+8.0
Box Nebula	Lupus	+10.5

Astrophysics is full of surprises. One would have thought, intuitively at least, that the larger the star, the greater its reserves of hydrogen are, and the longer it will live. In fact, the opposite is true. The largest stars have the shortest lives. A star half the mass of the Sun will remain on the main sequence for about 100 billion years, while stars just three times larger than the Sun spend only 300 million years there.

Stars under about 8 solar masses end their lives as planetary nebulae, but the small percentage of stars harboring masses greater than this end their lives in a final cataclysmic fireworks display. High mass stars tend to be hot and energetic, squandering

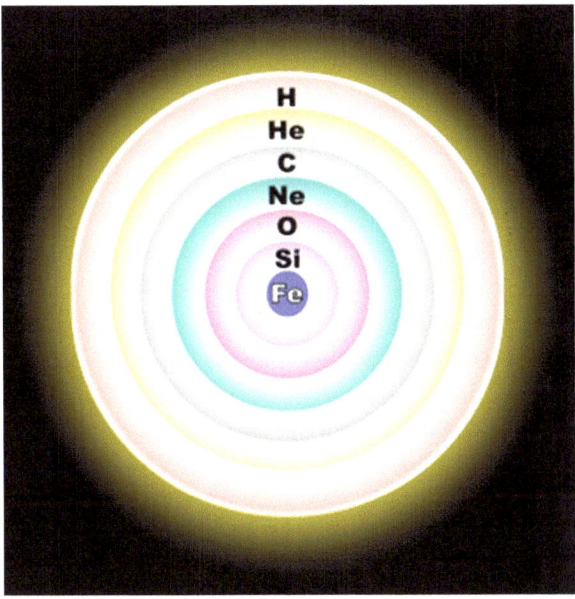

Fig. 7.11 Onion layer structure of a highly evolved massive star (Image © R. J. Hall. Used with permission)

their reserves of hydrogen and converting it into helium and thus become red super giants. You can see several bright red supergiant stars in the night sky. Betelgeuse, in the constellation of Orion, and Antares, in Scorpio, are good examples. Red supergiants burn helium into carbon and oxygen as lower-mass stars do, but the former then undergo further rounds of nuclear fusion (often in layers above the core, see below), creating still heavier atomic nuclei, such as magnesium, sulfur and silicon.

However, something strange happens after the star synthesizes silicon—it sounds the death knell for these beautiful giant stars. Silicon gets converted into iron, and iron does not burn (in a nuclear sense) at all. Because the synthesis of nuclei with masses greater than iron requires more energy than they release by undergoing fusion, an energy crisis ensues; the star collapses under gravity and implodes, becoming a *supernova*, shining briefly with the luminosity of a billion stars.

Only the core of the star remains after its violent death. The electrons and protons making up the matter in the supernova remnant combine to form neutrons, releasing a torrent of nearly mass-less particles called neutrinos in the process. The remnant is like a giant atomic nucleus—a *neutron star*. And just as a pirouetting skater spins more rapidly as she draws her arms inward, the tiny stellar core—only a few tens of kilometers across—speeds up its rate of rotation. Although stars like the Sun complete one revolution on their axis in weeks, neutron stars have revolution periods of only fractions of a second. These bizarre stellar corpses are what astronomers called pulsars, the fastest of which rotate at speeds approaching 10 % the speed of light!

Neutron stars can't be any old size, though. There is a limit to the amount of matter a neutron star can possess before gravity gets the upper hand. If the burnt-out core exceeds about two solar masses, then nothing can prevent gravity from crushing it out of existence. The star is doomed to end its life as a black hole—a place where a star once was.

Observing a Supernova Remnant

One night, back when the civilization of Egypt was in its infancy, a massive star ended its life in cataclysmic violence, shedding its evolved matter to the dark void of interstellar space in a stupendous explosion we now call a Type II Supernova. To our distant ancestors, gazing towards the constellation of Cygnus the Swan, the sight must have filled them with fear and wonder, as one of the brighter stars in the night sky became easily visible by day, slowly fading into obscurity over the passing weeks and months. Today, 5,000 years later, it is remarkable that we can still see the remnants of this rare and extraordinary event—the mysterious **Veil Nebula**.

Tucked away among the background of stars of the Cygnus Milky Way, the Veil is faint but very extensive. From a dark rural sky, its brightest structures can be seen in binoculars or better still, a small, rich field telescope. Looking towards the celestial Swan's outstretched wings, first locate Eta Cygni and move a few degrees southward until you to get the fourth magnitude star 52 Cygni. Careful inspection of the sky immediately north and south of 52 Cygni using 15×70 binoculars reveal the tell-tale signs of the western section of the Veil—designated **NGC 6960**—but you can obtain a much better view using a 3-in. (75-mm) or larger telescope and a low-power wide-angle eyepiece.

To enhance the view still further, insert an OIII filter, which greatly increases the contrast between the background sky and the Veil, making its structure stand out prominently. Indeed, an OIII filter is an absolute necessity if you plan to observe from even moderately dark skies. Such a filter will dim the background stars by about one stellar magnitude but greatly enhance the nebula's visibility at the eyepiece. Using this approach, in the $3°$ field of a 28× eyepiece, a 4-in. refractor gives a superb view of the entire structure. The Eastern Veil—**NGC 6992**—is undoubtedly easier to see than its western counterpart, in part because it doesn't live in the glare of a bright foreground star like 52 Cygni.

Once you've located these structures in your low-power field, cast your gaze towards the region of sky between the eastern and western sections of the Veil. Careful inspection with a 4-in. or larger 'scope will show up a third section some $1.5°$ north-northeast of 52 Cygni—a faint wedge of nebulosity—known as **Pickering's Triangular Wisp**, spanning an area of sky roughly equivalent to that of the full Moon. In its entirety, the Veil encompasses a vast prairie of sky about 9 square degrees in extent. That alone is ample evidence that the structure is thousands of years old. If you have doubts, turn your telescope to the Crab Nebula (M1) in Taurus. Having exploded just over 1,000 years ago (local time), the Crab Nebula covers only 24 square arc minutes, making the Veil over 1,300 times larger!

Centering the star 52 Cygni in the field of view and increasing the power to 100× in the 4-in. enables me to better delineate the finer details of the western Veil. In the 0.8° field served up by the eyepiece, you can follow the winding course of NGC 6960, like smoke issuing from a chimney on a still night. The structure is about 1° long and at most, about a fifth of a lunar diameter in width. The same exercise can be done with NGC 6992 that, in places, appears about twice as wide as NGC 6960.

Though smaller instruments are ideal for framing the entire Veil, they lack both the light grasp and resolution to discern any fine structure from this bright super-nova remnant. Pointing a 10-in. F/5 Dob towards the Veil and an OIII filter installed, the view is utterly transformed. What smaller instruments can only hint at, the 10-in. makes obvious—innumerable thin, braided wisps representing the incandescent shock fronts of the death star can be seen lacing their way through both the eastern and western sections of the nebula. Indeed, the brightest sections take on the appearance of some long exposure images taken with small refractors. Unlike those lovely images, of course, the view through the eyepiece is decidedly monochrome (you might pick up a faint greenish glow though). Experience shows that instruments with apertures upwards of 16 in. and a crystal clear sky are neces-sary to have any chance of seeing even the most garish colors of the Veil Nebula. And if you ever get an opportunity to look at this "legacy to star death" through a very large telescope from a dark sky site, don't miss it. It will change the way you view the heavens forever!

Activity: Observe the Crab Nebula (M1) in Taurus

In AD 1054, a brilliant new star shining four times brighter than the planet Venus appeared near Taurus the Bull's southern horn. For more than 3 weeks it remained visible during daylight hours, taking more than a year to fade from view. Today, we know this supernova remnant as the **Crab Nebula** (**M1**). The Anglo-Irish astrono-mer, William Parsons, the Third Earl of Rosse, sketched M1 in 1844 as it looked in his 72-in. reflecting telescope. Other astronomers noted the object's resemblance to a crab, and the name evidently stuck.

You can find M1 by first centering the fourth magnitude star Zeta Tauri in the field of view and then pan 1° to the northwest. An 80-mm 'scope can easily reveal its strong oval shape, one and a half times longer than it is wide and orientated northwest to southeast.

That brings our brief exploration of stellar evolution to an end. Time to look more closely at the great variety of stars that inhabit the heavenly realms.

Fig. 7.12 The magnificent Crab Nebula (Image courtesy of NASA)

A Sky Full of Double Stars

To the unaided eye, the fixed stars appear solitary, lonely objects. But the majority of them actually possess one or more companions, either too faint or too close to a brighter member to be seen without telescopic aid. Many of these systems are achingly beautiful, their constituent stars being of a different hue, creating lovely contrasting colors when examined in the telescope.

There are essentially two types of double star: true binaries, those that are physically associated with each other and optical doubles, and those that only give the appearance of such, being situated along more or less the same line of sight but at different distances from the Solar System.

True binary and multiple-star systems are gravitationally bound up with each other, with all components orbiting around a point marked by their common center of mass (barycenter). And though they are constantly in motion, the components of binary and multiple-star systems cannot be seen to move in the telescope, owing to their enormous distances, but by measuring their angular separations and position relative to the brightest member (primary) over years and centuries, the semblance of an orbit can be traced out. In this way, amateur astronomers are able to observe a canonical demonstration of the application of Newtonian mechanics beyond the Solar System. How cool is that?

Fig. 7.13 The incomparably beautiful Albireo in Cygnus (Image © Mike Pearson. Used with permission)

Many of the finest doubles in the heavens were discovered in the golden age of visual astronomy, when folk of extraordinary dedication and skill turned their long, classical refractors toward them systematically, recording their tiny motions over many human generations.

The closest star to the Sun, Alpha Centauri, is a triple system consisting of a brighter binary pair and a much fainter red dwarf, Proxima, slightly closer than Alpha is to our Sun.

Activity: Observe a Beautiful Double Star System This Evening

Go out of a dark evening under a transparent sky and locate the middle star in the familiar Big Dipper, Mizar. Can you see its fourth magnitude companion, Alcor, with your unaided eye? Now point your telescope at the stars, employing a magnification of about 50×. You should see that Mizar itself is double, of equal brightness, in fact, to Alcor. This companion was first seen by the Italian astronomer Giovanni Riccioli in 1650, making Mizar the first double star to be discovered in the telescopic age. Here are some famous double- and multiple-star systems.

Star	Constellation	Comments
Albireo	Cygnus	Striking orange and green components
Algieba	Leo	A nice close pair in orange/yellow
Castor	Gemini	Lovely high magnification pair
Cor Caroli	Canes Venatici	Splendidly easy in white and lilac
Epsilon Lyrae	Lyra	The famous Double Double. Must-see sight!
Almach	Andromeda	Striking color contrast pair!

Double stars are arguably one of the best ways of testing both your telescope's optics and the prevailing conditions under which you observe. There is no universally acceptable rule to establish how far two stars are apart before they can be separated. In the nineteenth century, observers such as the Reverend William Rutter Dawes (1799–1868) produced an empirical formula that serves as a guide to splitting double stars of equal brightness. After using a variety of telescopes (mainly refractors) of different aperture, Dawes arrived at an empirical formula that strictly applies to stars of equal brightness and of the sixth magnitude.

Dawes limit (arc seconds)=4.57/D, where D is the diameter of the telescope in inches.

Thus, according to the above formula, a 4-in. telescope out to resolve a pair of equally bright stars separated by a mere 1.14. Things get more complicated when one star is brighter than the other. The bigger the difference in brightness between the pairs, the harder it becomes to separate them. Attempts at modifying the Dawes formula to include stars of unequal brightness have not yielded any universally accepted 'rule'. That's not surprising, given the differences between telescopes and the enormous range of human visual acuity.

For example, this author has learned that color blind individuals may be able to detect fainter objects than 'normal' individuals, and they might, in addition, be able to resolve closer pairs than the average eye. Attempting to impose a hard and fast rule for everyone may be an exercise in futility.

Three parameters describe the telescopic appearance of a double star—the brightness of the stars involved, their angular separation and their relative orientation, known as their position angle (PA). Separation is measured in seconds of arc, whereas position angle is measured in angular degrees counterclockwise from north (denoted as 0°), with east being at 90°, south at 180° and west at 270°.

Many double stars are too close to be resolved with most ground-based telescopes but can be deduced by an examination of their spectra. For example Castor, a celebrated double star in Gemini, has two well resolved companions that orbit their common center of mass in about 500 years. Yet, each of these stars has yet another spectroscopic companion that whirls around it in just a few days.

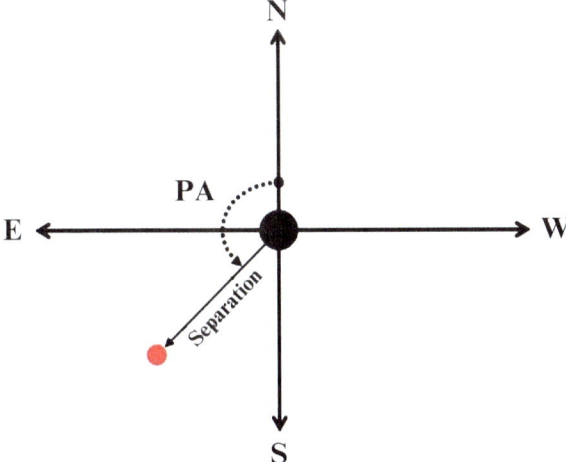

Fig. 7.14 Double stars are characterized by the angular separation (in arc seconds) from its primary and its position angle (in angular degrees measured counterclockwise from north). East and west are reversed in the above image (Image © Marc McClelland. Used with permission)

Activity: Discovering the Number and Diversity of Double Stars in Your Local Skies

Go out this evening with your grab 'n' go telescope and randomly select 20 stars in the sky. Examine each star at low and high power and record the number of such stars that appear to possess either close companions or members that are located far away. Repeat this procedure on several other samples of stars chosen randomly. Collate the information. How many stars appear to be binary or multiple?

Bessel's Star: 61 Cygni

Go outside tonight and locate the brightest star in Cygnus, the magnificent blue–white Deneb. Now, using ordinary 10×50 binoculars, pan about two binocular fields southeast. Chances are you'll come across a pair of golden suns, separated by a sliver of dark sky. This is the famous **61 Cygni** system. You'll get a much better view if you use a small, rich field telescope. A 3.1-in. (80-mm) refractor provides a splendid view at 30×, and it's even better at 96×, the brighter orange star shining with magnitude +5.2 with its fainter +6.1 companion displaced only 27 arc seconds to the southeast. These form a true binary system with an orbital period of about 700 years.

The orange colors of both components are striking, especially against the backdrop of fainter blue and white Milky Way stars. Indeed it is immediately clear that these are cooler stars (spectral class K) than our own Sun. From our comfortable vantage, the pair look relaxed, even serene. But careful inspection of this system over decades and centuries reveals that 61 Cygni is not an ordinary 'fixed star' but is sprinting across the sky.

That much became clear to the Italian astronomer Giuseppe Piazzi, based at Palermo Observatory, Sicily, as early as 1792, when he estimated that 61 Cygni was changing its position relative the background stars by as much as 5.2 arc seconds per year. Although that doesn't sound like much—about one-eighth of the apparent diameter of Jupiter at opposition—it was enormous by the standards of anything that was observed before.

Palermo, however, might as well have been a million miles away from the epicenter of astronomical research in Europe. As a result, Piazzi's observations went largely unnoticed for over a decade until 61 Cygni's extraordinary sojourns were again noticed by the German astronomer Friedrich Bessel, who published a report of the system's large proper motion in 1812. To Bessel, that was a sure sign that these golden suns were relatively nearby, but proving it was quite another matter.

Another astronomer, Wilhelm Struve, director of the Dorpat Observatory in Russia, provided the impetus for Bessel's groundbreaking work. If a star is truly nearby, Struve reasoned, it ought to possess one or more of the following characteristics: it should be fairly bright (nearer stars look brighter), have a large proper motion, and if it happens to be a binary star system, the two components ought to appear widely separated in comparison to the time it takes them to orbit each other. Struve agreed with Bessel that 61 Cygni was an excellent candidate to measure stellar distance. The method to be employed was trigonometric parallax. If a star is close, it should shift its position back and forth against the background stars in the sky as Earth orbits the Sun.

Bessel was fortunate enough to come of age in an era where astronomical telescopes, especially refractors, were being fashioned to unprecedented standards of accuracy and precision. His compatriot, Joseph Fraunhofer (1787–1826), used his optical genius to create large achromatic refractors on driven equatorial mounts—an absolute necessity for accurate positional measurements of stars to be undertaken. Thus, in a humble observatory in Konigsberg, Germany, Bessel had installed a purpose-built instrument called a heliometer to measure the parallax of 61 Cygni. Built by Fraunhofer, it consisted of a 6.5-in. (16-cm) object glass cut down the middle to create two semicircular halves. Each 'half' objective was separately mounted in such a way that one could be moved independently of the other. When perfectly aligned, the two half objectives form a single image, but as one half is moved relative to the other, two separate images are created. The amount of movement needed to superimpose the displaced images can be used to measure the angular separation between two or more objects. Using this method, Bessel measured background stars together with the brighter member of 61 Cygni to deduce the parallax of the star system.

Over a four-year period beginning in 1834, Bessel subjected 61 Cygni to intense scrutiny, repeating his measurements at least 16 times every night and many more times during nights of exceptional seeing. Conditions could sometimes be cruel, working as he did with his bare hands in an unheated observatory during freezing nights. But he persevered where many others would have given up. His results produced a parallax of 0.3483 arc seconds—only 10 % less than the modern accepted value—and corresponding to a distance of just over 10 light years.

For the first time in history, someone figured out the immense distance to a star—distances well beyond ordinary human understanding. In recognition of Bessel's work, John Herschel, then president of the Royal Astronomical Society, reminded his fellows that they had lived to see the day when the "sounding line in the Universe had a last touched bottom." It was, he continued, "the greatest and most glorious triumph which practical astronomy had ever witnessed."

Observations of the 61 Cygni system continued throughout the twentieth century. In the 1960s, astronomers detected a slight wobble in the motion of these stars, suggesting the presence of a third unseen companion. Subsequent studies have since refuted that idea, although astronomers continue to monitor the system. Being K dwarves, the stars of 61 Cygni have very strong magnetic fields and are thus prone to frequent flaring; something that might not bode well for any life forms evolving on any attending planets. I wonder what secrets await further scrutiny of this magnificent system as the twenty first century unravels.

Some Challenging Pairs for Smaller Telescopes

The following list of stars is good targets for smaller telescopes, requiring good optics and good conditions in equal measure to see them at their best. Try your hand at one or more of these as they appear in their appointed season.

Star	Constellation	Comments
Delta Cygni	Cygnus	Tough for 10-cm apertures and less
Theta Aurigae	Aurigae	Tough for 10-cm apertures and less
Rigel	Orion	Tough for a 6-cm aperture
Iota Cassiopeaie	Cassiopeia (triple)	Challenge for an 8-cm aperture
Mu Cygni	Cygnus	Challenge for 10-cm aperture
Eta Geminorum	Gemini	Tough target for a 10-cm glass
Zeta Aquarii	Aquarius	Challenge for a 6-cm glass
Pi Aquilae	Aquila	Challenge for a 12-cm aperture

**Activity: Determining the Minimum Aperture Required
to Split the Lyra Double Double**

Locate Epsilon 1 and Epsilon 2 Lyrae. Crank up the magnification of your telescope to about 150× until you can see both Epsilon 1 and Epsilon 2 as clearly double. Having prepared a series of cardboard aperture masks, explore what minimum aperture is required to clearly make up the four stars as distinct. Try decreasing your aperture in increments of 10 mm. Generally, somewhere between 50 and 60 mm are necessary, although your mileage may vary.

This brings our brief consideration of the world of double stars to an end. We shall be returning to many more of these objects in our summer and autumn tours of the heavens. Of course, you can always indulge your interest in double stars by consulting any good star atlas, which will have such pairs marked out for you. Most every constellation has one or more good double-star targets to investigate with your grab 'n' go telescope, allowing you to build up an impressive suite of observations in the field. Good luck on your journey.

Chapter 8

The Spring Galaxies

We inhabit an expanding universe. But it's not expanding at any old rate. If it did so more slowly than it does, then all the matter within it would be subject to the relentless grip of gravity, and we'd wind up with a cosmos littered only with black holes. If, on the other hand, the expansion was more rapid, then gravity's grip would have no teeth, and all we'd see is an amorphous cloud of gas and dust. Only by expanding at its current rate—neither too fast nor too slow—can a universe with stars, planets and living creatures co-exist.

Thank goodness for small mercies!

Galaxies are the largest structures visible to the human eye. In the scheme of things, they represent just a few percent of all the matter that exists, yet without their teeming numbers, some one trillion strong, there would be no life and nothing to contemplate. And though they are faint and fuzzy in even the largest backyard 'scope, they embody a kind of Mecca for the contemplative deep sky observer.

Spring is an excellent time to embark on a study of the galaxies. We have already discussed much of the real estate that makes up our own galaxy, the Milky Way. We have discussed and explored the multitude of stars and the gaseous nebulae out of which they were formed. We have explored star birth, mid-life and death. But there is yet another type of cluster associated with galaxies that is in many ways more mysterious than all the rest—the globular cluster.

Activity: Observe M13 in Hercules

Forget galaxies for the moment. High in the east on spring evenings, the large and sprawling constellation of Hercules is ripe for observation with your grab 'n' go telescope. Locate the famous Keystone—a prominent trapezium-shaped structure

N. English, *Grab 'n' Go Astronomy*, The Patrick Moore Practical Astronomy Series,   137
DOI 10.1007/978-1-4939-0826-4_8, © Springer Science+Business Media New York 2014

Fig. 8.1 Structure of the Milky Way Galaxy, showing the position of the Sun and its major spiral arms (Image courtesy of NASA)

at the heart of the constellation. Imagine a line drawn between Eta and Zeta Herculis. Starting at Eta, move to a spot roughly one-third of the way from Eta to Zeta. Here a low power eyepiece in an 80-mm refractor will show you the fuzzy outline of M13, the great globular cluster in Hercules.

To get the best look at this magnitude +5.7 object, you'll need to increase the power from 80× to 150× in a small telescope. An 80-mm refractor at 83× does a nice job bringing out this bauble of starlight from the background sky. About 300,000 stars are concentrated into a spherical structure with an apparent size roughly similar to that of the full Moon. Only a few of the outermost stars in the cluster can be faithfully resolved with such a small apertures, but a 5-in. (127-mm) telescope improves the view dramatically with hundreds of stars cleanly resolved to almost the center. M13 is just one of several hundred clusters that form a spherical halo around our galaxy and other galaxies. The distribution of such globular clusters—most concentrated around the center of our galaxy was used by the American astronomer Harlow Shapley to deduce its gross structure.

Why is the northern winter sky almost bereft of globular clusters?

Fig. 8.2 The great globular cluster in Hercules (Image © Mike Pearson. Used with permission)

This is due to our vantage point during winter compared with our summer view. Because globular clusters are concentrated in a direction towards the Galactic center, they are observed in much greater numbers than in winter, when our sights are directed towards intergalactic space. These simple observations help us to work out our place in space.

The Milky Way is a barred spiral galaxy some 100,000–120,000 light years in diameter and contains approximately 100–400 billion stars. It may contain at least as many planets as well. The Solar System is located within the disk, about 27,000 light years away from the Galactic center on the inner edge of a spiral-shaped concentration of gas and dust called the Orion-Cygnus Arm. The stars in the inner $\approx$ 10,000 light years form a bulge and one or more bars that radiate from the bulge. The very center is marked by an intense radio source named Sagittarius A, which is likely to be a super-massive black hole. Indeed, all galaxies harbor super-massive black holes at their centers. Furthermore, the size of the black hole is strongly correlated to the overall size of the galaxy. Why this should be the case is still unclear.

Galaxies are vast collections of stars, gas and interstellar dust. All stars, so far as we know, are derived from within them. The population of stars in galaxies varies considerably, from a few hundred thousand to over a trillion. Galaxies also exhibit great diversity of morphology, but astronomers have managed to classify them into a small number of categories. Broadly speaking, three major classes are known.

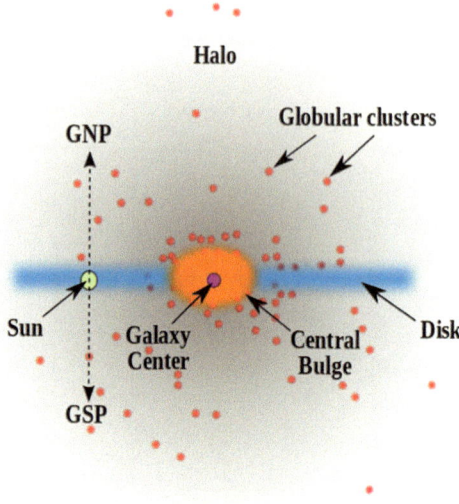

Fig. 8.3 Side-on view of the Milky Way showing the main components (Image © Marc McClelland. Used with permission)

Spirals

These appear as flat white disks, with yellowish bulges at their centers and beautiful spiral arms extending outwards from their nuclei. Both old and young stellar populations are found in spirals.

Ellipticals

These are more rounded or spherical in shape and have far more dust that make them appear redder. Most of the stars in these galaxies are old and highly evolved.

Irregulars

Display shapes that are neither spiral nor elliptical. Irregular galaxies have young and old stellar populations. These galaxies are thought to have come about from mergers between galaxies.

The American astronomer Edwin P. Hubble was the first person to put the many observed types of galaxy into some sort of order. Known as the Hubble

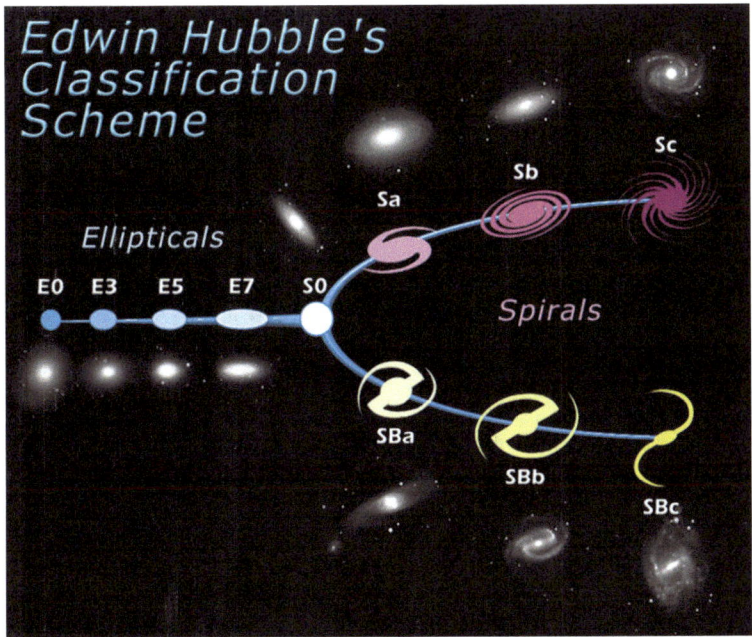

Fig. 8.4 Hubble's Tuning Fork diagram of galaxy classification (Image courtesy of NASA/HST Heritage Project)

classification scheme, it provides a means of categorizing the wide variety of elliptical and spiral galaxies based on his famous Tuning Fork diagram.

Spirals are subdivided into:

Sa—tightly wound, smooth arms; large, bright central bulge

Sb—less tightly wound spiral arms than Sa; somewhat fainter bulge

Sc—loosely wound spiral arms, clearly resolved into individual stellar clusters and nebulae; smaller, fainter bulge

Other spiral galaxies appear to have a prominent bar and are thus classified SBa, SBb and SBc, respectively.

Similarly, elliptical galaxies are divided into seven sub-classes (E1–E7). All of this appears a bit academic and you'd be right in thinking it only really matters to professional astronomers engaged in galactic research.

Galaxy Groupings

Our galaxy, the Milky Way, is a barred spiral of almost perfect symmetry. But it is only one of a larger consortium of island universes known as the Local Group, which comprises a 54-strong cohort of galaxies within a sphere roughly 10 million light years in diameter.

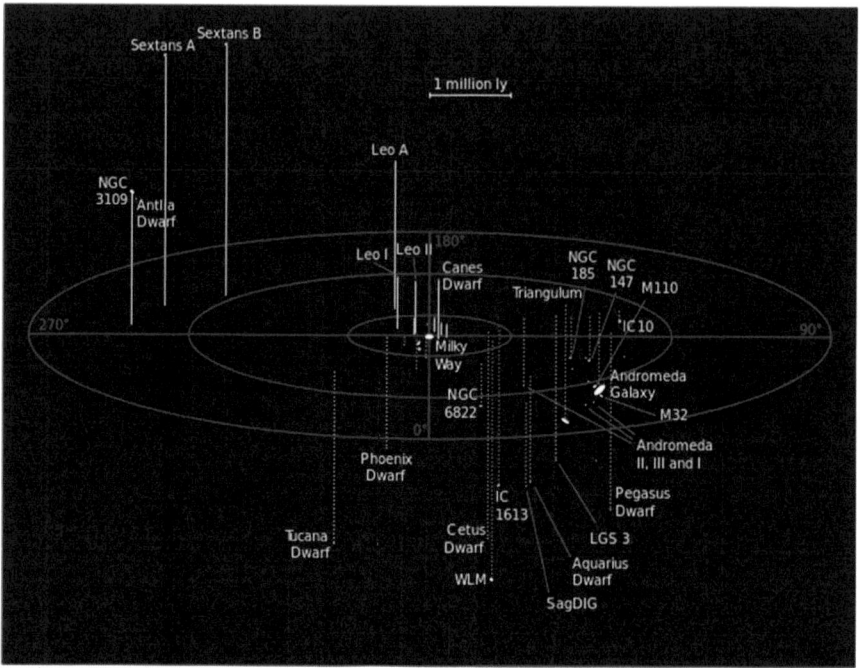

Fig. 8.5 Schematic of the Local Group (Image courtesy of NASA)

Prominent members of the Local Group include the Andromeda and Triangulum galaxies (explored later), as well as many so-called dwarf galaxies, including the Small and Large Magellanic Clouds, best seen from far southerly latitudes.

Further out from the Local Group, we soon encounter other galaxy groupings, and here we'll take a look at two galaxies easily visible in small grab 'n' go telescopes from a dark sky.

Activity: Observe a Galactic Odd Couple

When darkness finally falls on the landscape during spring evenings, cast your sights on a beautiful galactic duo located high in the northern sky among the barren reaches of Ursa Major. Our magnificent galactic neighbors **M81** and **M82** can be easily spotted in 10×50 binoculars by following the diagonal line from Phecda to Dubhe in the 'bowl' of the constellation westward until you hit the fourth magnitude star 24 Ursae Majoris. If you place that star toward the western edge of your binocular view, the seventh and eighth magnitude M81 and M82 should show up towards the center of your field. You'll need the help of more magnification and a little greater aperture to get a more immersive view, though. With an 80-mm f/9 refractor, the pair is beautifully framed in a low power eyepiece. M81, the more southerly of the two, is noticeably brighter and more extended than M82, which in contrast, looks like a sliver of light.

The views of both galaxies improve still more in a 4-in. refractor at 61×, where M81 is transformed into a very distinct oval some 20 arc minutes across (about two-thirds of a lunar diameter), becoming gradually fainter as one leaves its center. This author finds cranking the magnification up to 125× greatly helps to darken the sky around M81, allowing hints of its spiral structure to be glimpsed with averted vision. At this magnification, it is easy to make out several foreground stars superimposed on the galactic disc. Viewing M81 in a 6-in. (152-mm) f/5 reflector clearly shows the face-on spiral with arms streaming out from the bright galactic core.

Turning our attention now to M82, the view could hardly be more different. Shining with an integrated magnitude of 8.4, you might expect it to be dramatically less conspicuous than its spiral neighbor, but ironically, it is much easier to delineate structure from it, even in small aperture 'scopes. In a 4-in. at 72×, M82 is transformed into a cigar-shaped object about 11 arc minutes long and about 4 arc minutes wide. Even at this moderate magnification, M82 shows a clear mottling in its edge-on structure. At 125× in a 4-in., M82 looks as though it is bifurcated by a prominent dark lane that 'separates' the galaxy into two distinct regions along its broad section. Excellent fine structure can be seen in 'scopes larger than 8 in. at high magnification. Indeed, it is arguably one of the easiest galaxies to 'unpick' using backyard 'scopes, and you can easily find yourself studying it for hours on end.

Chalk and Cheese

Despite their relatively close proximity to our own galaxy—about 8 million light years according to current estimates—these galaxies couldn't be more different. M81 is a large spiral galaxy, much like our own Milky Way, containing a few hundred billion stars. M82 is a small, highly irregular galaxy with perhaps a few tens of billions of suns. Detailed observations made at visible, infrared and radio wavelengths point to a catastrophic past for M82. Specifically, astronomers speculate that an enormous explosion ripped through the galaxy's nucleus, sending shock waves far out into its outer extremities. Intriguingly, no one knows just how or why such a catastrophic event took place.

Not only are they located at more or less the same distance from us, but these galaxies are also physically associated, too, with only 100,000 light years or so separating them. That may sound like a huge distance in the scheme of things, but in reality it brings the galaxies some twenty times closer than the Andromeda Galaxy (M31) is to our own Milky Way. From a planet orbiting a star in M82, the spiral structure of M81 would dominate the night sky with mind-boggling levels of detail available even to the unaided eye. Messier 81 is the largest galaxy in the so-called M81 Group, an assortment of about 35 galaxies located in the constellation Ursa Major. The distance from Earth to the group is approximately 12 million light years, making this one of the closest groups to our own galactic club, the Local Group, which contains our own Milky Way. M81 is gravitationally interacting with M81 and its smaller neighbor **NGC3077.** The interactions have stripped some hydrogen gas away from all three galaxies, leading to the formation of filamentary gas structures in the group. Moreover, the interactions have also caused some interstellar gas to fall into M82 and NGC 3077, which is currently causing a wave of star formation within the centers of these two galaxies.

Going Deeper

Spring evenings are arguably the best time to explore an even larger galaxy grouping spread out across Virgo and Coma Berenices. The Virgo cluster is a massive cluster of galaxies that dominates the so-called Virgo Supercluster. There are roughly 2,000 galaxies in this cluster (although 90 % of them are dwarf galaxies). This cluster has a diameter of approximately 15 million light years, which is not much larger than our Local Group, but it contains 50 times the number of galaxies!

In this section, we'll pick out a few of the more prominent galaxies from this supercluster, accessible to a small backyard telescope. In general, a low power eyepiece is used for locating and centering the galaxies, while a medium to high power eyepiece is beneficial for extracting detail from the image at the precipice of visibility.

M59 and M60

These are a pair of elliptical galaxies, easily picked up in binoculars, though much more distinctly seen in a small telescope using higher magnification (50× or above). Finding this galaxy pair requires a good measure of patience and a low power eyepiece. Start by locating the bright Vindemiatrix (Epsilon Virginis) almost due east of Denebola. Then, star hop 4.5° west and a shade north of Epsilon to locate one of the largest elliptical galaxies presently known—**M60**, shining with an integrated magnitude of +8.8. Also in the field to the west (the direction of drift) is the other Messier we're looking for, the elliptical galaxy **M59**, a full magnitude dimmer than its neighbor. Both M59 and M60 are located at enormous distances—some 55 million light years distant. That's quite something to contemplate when you center both galaxies in the same field of view!

Question: Can you make out their elliptical structure?

M58

This is a magnitude +9.6 barred spiral galaxy located about 2° northwest of fifth magnitude Rho Virginis. At low power, you can make **M58's** star-like nucleus, surrounded by a faint halo. Cranking up the power to 80× or so shows the mottled halo and some suggestions of structure. Averted vision may allow you to glimpse the condensed bar running through its center.

M89 and M90

This duo represent an excellent study in contrasts, for this galactic pairing show an elliptical galaxy (**M89**) and a spiral galaxy (**M90**) side by side, in the same low power field of view, with a mere 40 arc minutes separating the two. Although both shine at roughly the same magnitude (+9.7), the spiral nature of M90 is generally thought to be more attractive to the eye. Twice as long as it is wide, M90 long axis is orientated northwest to southeast. Using an 80-mm refractor with a power of 83×, you can begin to divine hints of mottling in its bright, elongated core. With dark-adapted eyes, you ought to easily see the more than subtle differences between it and the somewhat more rounded M89.

M91

Actually located across the border in Coma Berenices, **M91** is an excellent example of a barred spiral galaxy, located about 4.5° northwest of magnitude +4.5 Rho Virginis. Moderate powers in an 80-mm refractor will show its general structure, and with averted vision you may be able to make out some outlying spiral structure and mottling within the core. High powers, patience and a dark sky are required in equal measure to delineate any structure in its spiral arms.

M88

Located at 4.5° southeast of the star 11 Comae, a low power eyepiece from a dark moonless sky reveals the spindle-like morphology of the magnitude +9.6 spiral. As a bonus observation, can you make out the two 12th magnitude stars, separated by about 30 s of arc kissing the southeastern-most tip of its spiral arms?

*Activity: Locate **M84** and **M86** in Virgo*

Without looking up the descriptions, observe their appearance in the telescope. Can you make out any differences between them? If so, what?

More to observe: There are many objects in the Virgo Super Cluster of galaxies that warrant finding and studying. It is unfortunately beyond the scope of this book to mention them all, but here are some other Messier spring galaxies well worth the effort to track down:

M87: Fairly bright (magnitude +8.6) elliptical galaxy in Virgo.
M49: An even brighter elliptical (+8.4) in Virgo.
M61: Faint (+9.6) but remarkable spiral galaxy in Virgo.
M99: Beautiful but enigmatic (+9.4) face-on spiral in Coma Berenices.
M98: A rather elongated spiral galaxy (edge on) in Coma.
M100: Faint but very remarkable face-on spiral in Coma.
M85: Fairly bright (+8.6) galaxy in Virgo that has a very clearly defined lenticular shape.

This brings our study of spring galaxies to an end. We will get another opportunity to study much nearer—and grander—island universes, when we visit the skies of autumn.

Chapter 9

A Study of Eclipses

For the vast majority of human history, the sight of a solar eclipse must have filled souls with wonder and terror in equal measure. The temperature drops, ole Aeolus stirs up strange breezes, the birds refrain from their song and many other animals either flee or cower in their dens.

A person in his or her lifetime can rightly expect to see about 50 lunar eclipses—more than half of them total. And of the Sun, she or he can reasonably hope to savor about 30 partial events. Totality, however, is an exceedingly rare event at any one location. Still, bewitched by a quirk of celestial mechanics, adventurous individuals are willing to travel half way around the world to witness a phenomenon that can only last 7.5 min at the very most.

Eclipse chasing is a way of life.

According to the Greek historian Herodotus, the ancient Greek astronomer, Thales of Miletus (624–546 BC) predicted an eclipse that occurred at the site of the battle of Halys, where the Medians fought the Lydians on May 28, 585 BC. As the eclipse approached totality, both armies laid down their arms and declared peace. In 480 BC Herodotus himself observed an eclipse first hand just as the Persian King, Xerxes, made final preparations for his failed campaign against the Greeks.

Half a world away, the ancient civilization of China cultivated the science of astronomy, recording solar eclipses as far back as 720 BC Court astronomers (or more accurately astrologers) were held in very high esteem by the Chinese aristocracy, where they often cultivated great power and authority. If they were cunning enough, they could pull the purse strings of their noble patrons for a lifetime. But other charlatans of predictive astronomy fared far worse. The Emperor Zhong Kang supposedly had two astrologers—Hsi and Ho—beheaded, when they failed to predict an eclipse that occurred during his reign. Slowly, over the centuries, Chinese astronomical knowledge grew, and by the fifth century BC they mastered the skill that the Greeks had mastered before them, by making better positional measurements of the Moon and Sun's motion on the celestial sphere.

Eclipses have proven invaluable in helping to date events in human history, too. For example, a solar eclipse occurring on June 15, 763 BC recorded in an ancient Assyrian text has proven seminal in piecing together the allegory of the Orient.

New Testament scholars, attempting to establish the exact date of Good Friday, have asserted that the darkness described during Jesus's crucifixion was attributed to a solar eclipse. This has not yielded conclusive results, however, owing to the paucity of reliable Western accounts before AD 800.

The first known telescopic observation of a total solar eclipse was made in France back in 1706. Within a decade, the English astronomer Edmund Halley, of comet fame, observed the solar eclipse of May 3, 1715, at Cambridge. By the mid-nineteenth century, scientific understanding of the Sun was improving through observations of the Sun's corona during solar eclipses. The corona was identified as part of the Sun's atmosphere in 1842, and the first photograph (or, more accurately, a daguerreotype) of a total solar eclipse was made on July 28, 1851. And spectroscopic observations were being conducted during a solar eclipse as early as 1868, helping to refine our knowledge of its chemical makeup.

Solar eclipses also proved indispensible in proving Einstein's epochal theory of General Relativity. Einstein had predicted that the position of the distant stars in the vicinity of the Sun would be very slightly altered owing to the bending of the fabric of space–time due to the Sun's mass. In 1919, the British astrophysicist Sir Arthur Eddington, ventured to the paradise island of Principe, off the west coast of Africa, in order to record the solar eclipse of May 29. His photographs during totality, upon closer scrutiny, vindicated Einstein's genius.

An eclipse occurs when one object in the sky moves into the shadow of another. The term is most often used to describe either a solar eclipse, when the Moon's shadow crosses Earth's surface, or a lunar eclipse, which occurs when the Moon moves into the shadow of our world. If you think about it for a few moments, you'll realize that for a true eclipse to occur, both participating bodies must lie in the same orbital plane.

The fact that we don't see eclipses, either of the Sun or Moon, that often, is proof enough that the Moon's orbit is inclined a little in comparison to both Earth and the Sun. Where they cross denotes the 'nodes' of its orbit. Thus, for example, as the Moon is traveling north, it intersects Earth's orbital plane at the ascending node. Conversely, when gray Luna moves southward it eventually crosses Earth's orbit at the so-called descending node. To get a lunar eclipse, our satellite has to be positioned at or very close to one of these nodes at full Moon phase. On the other hand, a solar eclipse always occurs at new Moon.

Owing to the cyclical nature of the coincidence of the orbital planes of the eclipsing bodies, we can predict when they will occur with astonishing accuracy. These cycles, or *saros*, occur approximately every 18 years and 11 days and represent the moment when Earth, the Sun and the Moon display the same geometric relation with respect to each other, as they did in the last eclipse. The only snag is that they will occur exactly one-third of the way around the planet from the previous one.

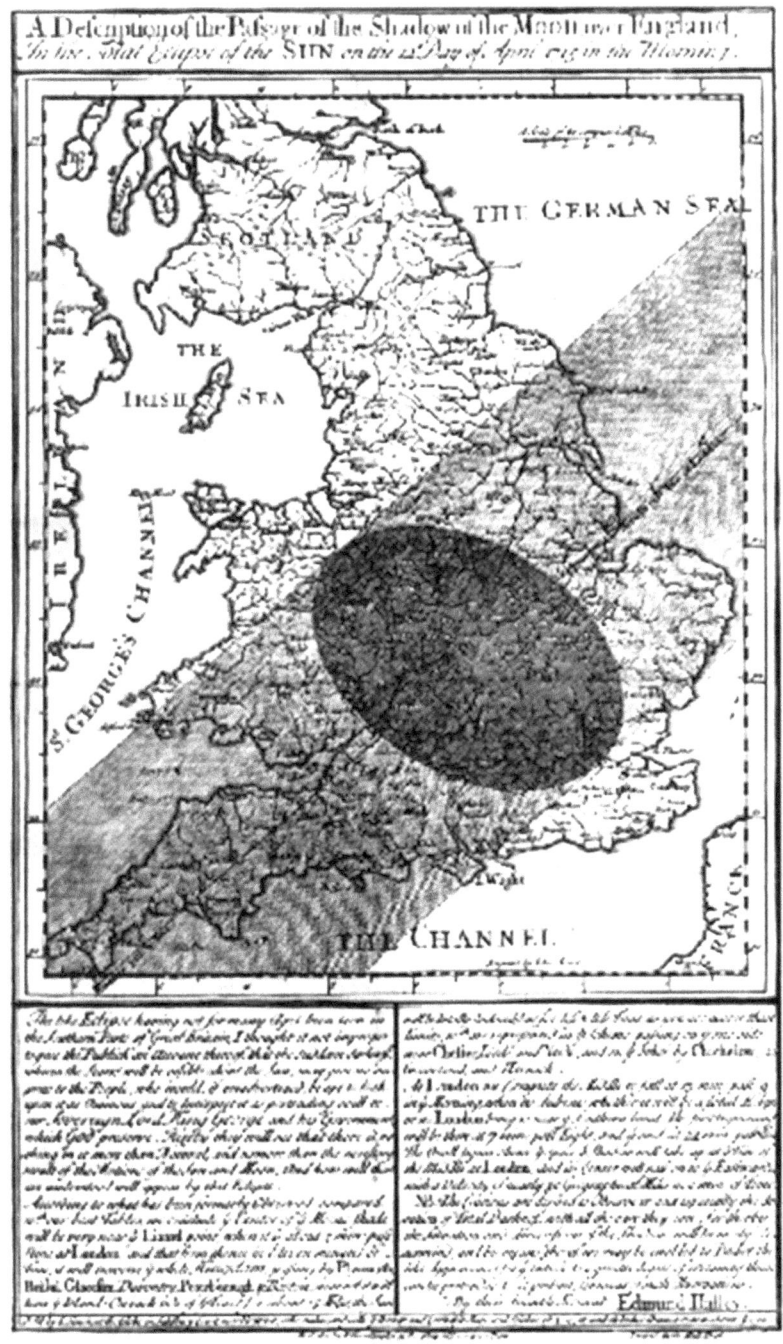

Fig. 9.1 A predictive map made by Sir Edmund Halley for the path of the Moon's umbral shadow (http://en.wikipedia.org/wiki/File:Solar_eclipse_1715May03_Halley_map.png)

Fig. 9.2 Totality, as recorded by Sir Arthur Eddington, on May 29, 1919 (http://en.wikipedia. org/wiki/File:1919_eclipse_positive.jpg)

Some interesting mathematical facts emerge from a study of the saros. For example, we can expect a total solar eclipse somewhere on Earth every 18 months or so. No more than five solar eclipses—total, annular or partial—can occur in any one calendar year. And when a solar eclipse occurs, it is always preceded by a lunar eclipse, separated by a 14- or15-day interval.

Here's a question: Can you observe (a) a total solar eclipse or (b) a total lunar eclipse from Earth's North or South Pole?

A Lunar Eclipse

Shortly after supper on the evening of March 3, 2007, this author set up an old 4-in. Tele Vue Genesis refractor on its sturdy Gibraltar mount in anticipation of the night's promised spectacle. As the full Moon rose in the east, its silvery countenance was almost overwhelming, drowning out the light of all but the brightest stars.

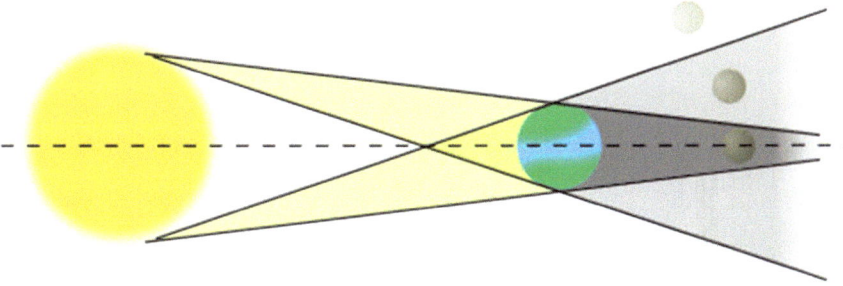

Fig. 9.3 The geometry of a lunar eclipse (Image © Marc McClelland. Used with permission)

Inserting a low power eyepiece, we watched as the eclipse commenced at about 8.30 p.m. local time. Not that anything looked untoward at first. That's because the initial phase of a lunar eclipse goes almost undetectable, as our satellite enters the Sun's penumbral shadow. By about quarter past nine, the eclipse took on a much more visually dramatic phase, as our satellite began the umbral phase of the eclipse.

We took our turns at the eyepiece as an eerie blackness—like one of those sinister interstellar clouds from the imagination of Sir Fred Hoyle—began to make its way across the lunar regolith, swallowing an ever increasing swathe of its vast globe. By half past ten, a most remarkable and surprising apparition unfolded—the lunar eclipse was now nearing totality. Yet, instead of completely enveloping it in darkness, a faint, cuprous glow appeared as if out of nowhere, on the dark side of the orb, becoming increasingly pervasive as the minutes passed. By 11:30, totality had arrived. We thought that the Moon looked more like a painting than any real thing. Rudely dark, its cuprous tones glowed like the dying embers of a furnace, with the lesser stars of Leo winking around it.

Such events are the life and soul of our hobby, for they allow us to become ringside spectators to the latent talent of Nature to surprise and delight. Most anyone can understand the origin of the creeping dark umbral phase, but a more curious person might ponder the peculiar rubicund glow during totality. The explanation lies with that thin veil of air—only 50 miles in extent—that blankets life on Earth from the lethal environs of space. If Earth had no atmosphere, the eclipse would never produce its characteristic coppery tones. That's because when the Sun, Moon and Earth are perfectly aligned, the terrestrial atmosphere generates its own 'aureole,' behaving as a lens of sorts, refracting and scattering blue and green rays more than red. The moon glows red for largely the same reasons the sky is blue—Rayleigh scattering.

No two lunar eclipses are ever the same. The vagaries of the weather seem to be one of the determining factors. Sky transparency also has a noticeable effect on the brightness and impact of a lunar eclipse. Clouds and moisture redden sunlight, which in turn makes the Moon appear more sanguine. Thin clouds can often veil the rich color of the eclipse at totality. And, every now and then, volcanic eruptions

Fig. 9.4 The lunar eclipse of March 3, 2007, arranged in a photographic montage (http://en.wikipedia.org/wiki/File:Lunar_eclipse_March_2007.jpg)

inject enormous quantities of ash into the atmosphere, where it is dissipated by winds. This can further darken the shadow of the eclipse, creating particularly strange effects. For example, in the aftermath of the Mount Pinatubo eruption in 1991, the penumbral lunar eclipse in July of the same year was seen to disappear completely, only rendered visible with the aid of binoculars.

Activity: During the next lunar eclipse, try to estimate the faintest stars that appear around the Moon as the event unravels.

The Glory of a Solar Eclipse

As spectacular as lunar eclipses can be, most people consider their solar counterparts to be far more compelling to watch. What is arguably most remarkable about a total solar eclipse is the exquisite symmetry of the event. Many people consider this extraordinary conjunction of stars and worlds to be simply miraculous.

Whether or not you regard it as such is not important, but the odds of such an event being replicated anywhere else in the cosmos must be virtually nil.

As we briefly touched on in a previous chapter discussing solar observing, *One should never look directly at the sun with or without an optical aid. Any attempt to do so can result in permanent eye damage or blindness.*

That said, there are a number of safe ways to observe the progress of a solar eclipse. By far the simplest is via pinhole projection. One can construct one's own pinhole, or, better still, look for natural pinholes in the leaves of trees that can often project an image of the solar disk onto the ground. If you wish to use optical aid, then Baader Astrosolar material (discussed previously) can be affixed to the objectives of small telescopes or binoculars to allow you to monitor the progress of the event from start to finish. Narrowband filters are especially useful during a total solar eclipse, as they can reveal spectacular prominences that are difficult if not well nigh impossible to observe in visible light.

Although partial solar eclipses are certainly spectacular, there is little in the way of science that can be done in the absence of totality, when the Sun vanishes behind the splendid orb of the new Moon. The most obvious change is the illumination of the landscape. The shadows cast by upright objects still remains but become far more diffuse. Color fades from the landscape, and everything is imparted with a dull, grayish hue. In the precious minutes prior to totality, the sky around the Sun manifests a unique shade of blue, inducing weird ruddy shadings on the horizon. Rarely, if the atmosphere is clear and tranquil, shadow bands—darker shadings interspersed by brighter diffuse regions—are seen to ripple their way across the surface of a whitewashed wall or smooth paper laid flat on the ground. If you happen to capture the event—which can only be a seen immediately before and after totality—then your experience of the eclipse will be greatly enriched. Photographers have found it quite a challenge to record them, either on film or with electronic eyes.

Of Beads and Diamonds

In the precious seconds just before totality, a keen-eyed observer might notice a series of bright 'beads' along the limb, culminating in a single bead—the famous diamond ring phenomenon. It was the English astronomer, Francis Baily (1774–1844), who was the first person to offer an explanation. Observing the total solar eclipse occurring on May 15, 1836, he correctly attributed the appearance of these beads to the irregular topography of the lunar surface, suggesting that the presence of upraised lands—hills and mountain ranges with their lofty peaks—allow sunlight to sliver through in some places but blocking it at others, hence *Baily's beads*.

Because the Moon's apparent size is smaller at apogee (its most distant point in its orbit from Earth) than that of the Sun, eclipses that occur when the Moon is in that orbital locus are not quite total, but 'annular,' displaying a ring of the solar photosphere uncovered.

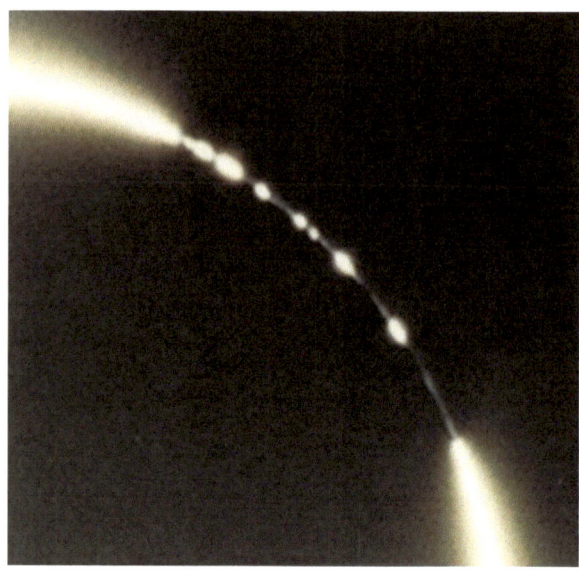

Fig. 9.5 Baily's beads (Image © Daniel Fischer. Used with permission)

Observed telescopically with an appropriate filter (discussed later in this book), the Sun displays a wealth of phenomena. Spectacular prominences are seen following the twisted magnetic field of our star. The chromosphere, a thin, ruddy layer of the Sun's atmosphere, is seen hugging its limb. But it is arguably the appearance of the vast solar corona—the Sun's outer atmosphere—that attracts the eye most strongly. Streaming their way outwards in delicate, wispy projections, during the most favorable conditions these structures can be traced out to distances as great as a few solar diameters. Astonishingly, the temperature of the corona—of the order of a million Kelvin—is much greater than the Sun's designated surface temperature of 5,800 K. It is thought that the solar magnetic field up-wells hotter material from its interior, funneling it outwards from the surface.

From the standpoint of chemistry, the corona is especially fascinating. The ultrahigh temperatures and strong magnetic field are in large part responsible for this. In the nineteenth century, spectroscopists discovered a peculiar species in the corona that matched none of the elements known on Earth. The mysterious substance was named coronium. Only in recent years have scientists been able to identify the substance as the highly ionized species of iron—Fe XIV—so hot that 14 of its outermost electrons were expelled from their orbits! That's plasma chemistry for you!

Many amateur astronomers find photographing solar eclipses to be very rewarding. But should the first-time observer invest those precious minutes of totality stuck behind a camera? Maybe those who have never before observed the magnificence of a solar eclipse would fare better by leaving their cameras at home!

Solar eclipses can also be seen from other worlds within our Solar System. For example, were we able to observe from Jupiter's cloudtops, five satellites—Amalthea, Io, Europa, Ganymede and Callisto—would be large enough to occult the Sun. The many smaller satellites further out, by virtue of their apparent size and more highly inclined orbits, can only be observed to transit the solar disk. Yet, during these Jovi-centric eclipses we, with our Earth-bound telescopes, are treated to a shadow transit—one of the most remarkable sights available to an amateur astronomer. Find out more about these transits in the chapter discussing Jupiter.

Historically, the phenomenon of satellite eclipses has been put to good use. For example, it was noticed early in telescopic history that their predicted times differed from the observed times in a repeatable manner. The first person to make a detailed study of this phenomenon was the Danish astronomer Ole Christensen Rømer (1644–1710), who, after studying the motions of Io for 8 years, correctly deduced

Fig. 9.6 Ole Rømer at work (http://en.wikipedia.org/wiki/File:Ole_R%C3%B8mer_at_work.jpg)

that these variations were caused by the changing distance between Jupiter and Earth as the two orbs raced around the Sun, causing the timing of the eclipse of Io to oscillate by up to 10 min earlier or later than predicted. It was left to others to infer the finite speed of light from Rømer's data. The Dutch astronomer, Christian Huygens, calculated that light waves move through the 'aether' at an astonishing pace—16.7 times Earth's diameter each and every second, or about 70 % of the currently accepted value!

Sadly, that brings to an end our brief study of eclipses. This author hopes you'll agree that they are among the most amazing of natural spectacles, more than worthy of attention and study, the life and blood of the grab 'n' go astronomer.

Chapter 10

Sol Invictus

One of the great misconceptions of modern astronomy concerns the idea that our Sun and the life-bearing world it has nurtured over eons of time, is in some sense 'common.' Diligent research across a dozen disciplines has cast severe doubt on that rather outdated and, dare it be said, naive notion. As was touched on in an earlier chapter, our Solar System was forged in a unique epoch in cosmic history, during the time when elements such as uranium and thorium reached their maximum abundance. Since that era, the creation of these elements via supernova explosions (the only known source of uranium and thorium) has lagged behind the rate at which these elements decay. Since this time, a little under 5 billion years ago, the availability of these elements for planet building has decreased.

The significance of this timing comes into sharp focus when one considers how crucial these radioactive heavy elements are to driving the vital process of plate tectonics on our planet. As luck would have it, the timing of Earth's formation and a number of subsequent events in the Sun ensured that enough uranium and thorium was present in the nascent Earth to sustain strong and enduring plate tectonics, creating the ideal ratio of continental to oceanic crust that has proven so critical to the recycling of nutrients and, when the time was most ripe, the emergence of a sentient technological civilization.

The Sun was not born with the characteristics we see today. Because of the decreased density of particles in its core, nuclear reactions were considerably more sluggish in the nascent Sun than they are now. Indeed even when the first life appeared on the planet some 3.8 billion years ago, it was still about 15 % fainter than it is in our epoch. The early onset of carbonate-silicate cycling in Earth's crust slowly removed the greenhouse gases to compensate for the Sun's increasing luminosity over the ages.

N. English, *Grab 'n' Go Astronomy*, The Patrick Moore Practical Astronomy Series, DOI 10.1007/978-1-4939-0826-4_10, © Springer Science+Business Media New York 2014

The Sun's location in the galaxy seems far from arbitrary. A number of professional astronomers have noted the exceptional chemical composition of the Sun's place of birth, particularly the abundance of a radioactive form of aluminum (aluminum 26) and its importance in purging the early Solar System of volatiles that would have left Earth with a dense, choking atmosphere and oceans far deeper and more hostile to life than the shallow oceans we are fortunate enough to have today. Only a unique cocktail of supernova events—occurring at the right times and at the right distances from the solar nebula—could ensure the delivery of the optimum chemistry for the emergence of Earth as a life-bearing world under the stewardship of the human species.

For over half a century, astronomers have searched the depths of our Milky Way Galaxy for a star that would qualify as a solar 'twin,' that is, showing all the characteristics necessary to sustain advanced life. The thinking in the astronomical community just a few decades ago was that given the galaxy's 200 billion stars and the great variety of differing properties for all those stars, it should not be too difficult to find many exact analogs to the Sun. However, in spite of diligent searches over several decades, astronomers have found only a handful of analogs and then some. You see, it is *much* easier to find twins of other stars but not so the Sun.

Atypical Mass

It is widely believed the Sun is a rather small star, dwarfed by many larger luminaries that decorate our night sky. In fact, the Sun is ranked in the top 4–8 % of the most massive stars in the Galaxy! Research over the last few decades has identified startling numbers of dwarf stars that are exceedingly faint and well beyond the means of a typical amateur to image or observe. In the distant past, it was even more bulky, having lost some 5–8 % of its mass over time.

Unique Chemical Composition?

Peruvian-born astronomer Jorge Meléndez, based at the Mount Stromlo Observatory, has spent the last few years comparing the abundances of different atomic elements in the Sun with abundances of the same elements seen in the most solar-similar stars to a precision never before achieved. This kind of differential element analysis found that the Sun seems to possess a unique elemental composition. In a series of papers published between 2006 and 2009 in the *Astrophysical Journal*, Meléndez compared some 20 of the very best solar analogs so far identified with the Sun and discovered that the latter exhibits a 20 % depletion of refractory elements relative to volatile elements. Refractory elements, which include calcium, iron and silicon, for example, comprise the bulk of the material making up interplanetary dust, meteoroids as well as the terrestrial planets. In contrast, the volatiles include the smallest and most abundant elements, such as oxygen, nitrogen and carbon, that factor into the gas molecules that make up the planetary atmospheres.

From his work so far Meléndez concludes that our kind of Solar System could be very rare indeed. Getting the right chemical inventory is quite tricky, with volatile-rich gas giants orbiting further out and rocky planets packed full of the right refractory elements orbiting closer to their stars.

Unusual Stability

Our Sun varies in luminosity by a mere 0.1 % over an 11-year cycle. What's more, photometric studies conducted on many sun-like stars show that this ultra low level of variability is the exception rather than the rule. Indeed recent research seems to indicate that our Sun could be the most stable middle-aged star known to astronomers.

General Location

As explained earlier, the Sun and its retinue of planets are located within the thin disk of our galaxy, oscillating with lower amplitudes above and below its mid-plane. It is situated at the co-rotation axis of the Milky Way, where its orbital period matches that of the neighboring spiral arms. Stars located inside or without this locus cross spiral arms more often and hence experience more frequent mass-extinction events.

More Sun Facts

The Sun is an almost perfectly spherical ball of hot plasma. Sporting a diameter of about 1.4 million km (865,374 miles); it is around 109 times the diameter of Earth and has a mass of 1.989×10^{30} kg, approximately 330,000 times the mass of Earth, which accounts for about 99.86 % of the total mass of the Solar System. Chemically, about three-quarters of the Sun's mass consists of hydrogen, while the rest is mostly helium. The remainder (1.69 %, which nonetheless equals 5,600 times the mass of Earth) consists of heavier elements, including oxygen, carbon, neon and iron, among others.

The Sun formed about 4.6 billion years ago from the gravitational collapse on the outskirts of a large molecular cloud. Most of the matter gathered in the center, while the rest flattened into an orbiting disk that would become the Solar System. The central mass became increasingly hot and dense, eventually initiating thermo-nuclear fusion in its core. Indeed, as we have seen, almost all stars form by this process. The Sun is classified as a G-type main-sequence star (G2V) based on spectral class, and it is informally designated as a yellow dwarf because its visible radiation is most intense in the yellow–green portion of the spectrum; and although it is actually white in color, from the surface of Earth it may appear yellow because of atmospheric scattering of blue light. In the spectral class label, *G2* indicates its surface temperature of approximately 5,778 K (5,505 °C), and *V* indicates that the Sun, like most stars, is a main sequence star and thus generates its energy by nuclear fusion of hydrogen nuclei into helium. In its core, the Sun fuses 620 million metric tons of hydrogen each second.

Once regarded by astronomers as a small and relatively insignificant star, the Sun is now thought to be brighter than about 85 % of the stars in the Milky Way Galaxy, about 70 % of which are now known to be red and brown dwarfs. The absolute magnitude of the Sun is +4.83; however, as the star closest to Earth, the Sun is the brightest object in the sky with an apparent magnitude of −26.74. The Sun's hot corona continuously expands in space, creating the solar wind, a stream of charged particles that extends to the *heliopause* at roughly 100 times further out than Earth's orbit. The bubble in the interstellar medium formed by the solar wind, the *heliosphere* is the largest continuous structure in the Solar System.

The Sun is currently traveling through the Local Interstellar Cloud (near to the G-cloud) in the Local Bubble zone, within the inner rim of the Orion Arm of the Milky Way Galaxy. Of the 50 nearest stellar systems within 17 light-years from Earth (the closest being a red dwarf named Proxima Centauri, at approximately 4.2 light-years away), the Sun ranks fourth in mass. The Sun orbits the center of the Milky Way at a distance of approximately 24,000–26,000 light years from the galactic center, completing one clockwise orbit, as viewed from the galactic north pole, in about 225–250 million years.

The mean distance of the Sun from Earth is approximately 1 astronomical unit (150,000,000 km, or 93,000,000 miles), though the distance varies as Earth moves from its closest (perihelion) in January to its furthest (aphelion) in July. At this average distance, light travels from the Sun to Earth in about 8 min and 19 s. But there is more to this last statement than meets the eye.

Amazingly, the radiation, which actually originates in the core of the Sun, takes about ten million years to zigzag its way to the surface (photosphere) before traveling through the vacuum of interplanetary space towards Earth. So, next time you turn your face towards the Sun on a cloudless afternoon, remember that this light is far more ancient than the human species!

The energy of this sunlight supports almost all life on Earth by photosynthesis and drives Earth's climate and weather. The enormous effect of the Sun on Earth has been recognized since prehistoric times, and the Sun has been regarded by some cultures as a deity. An accurate scientific understanding of the Sun developed slowly, and as recently as the nineteenth century prominent scientists had little knowledge of the Sun's physical composition and source of energy. This understanding is still developing, for there are a number of present-day anomalies in the Sun's behavior that remain unexplained.

The Sun is a Population I, or heavy-element-rich, star. Indeed, for a star of its age, it is the most enriched in heavier elements of all stars known. The formation of the Sun may have been triggered by shockwaves from one or more nearby supernovae. This is suggested by a high abundance of heavy elements in the Solar

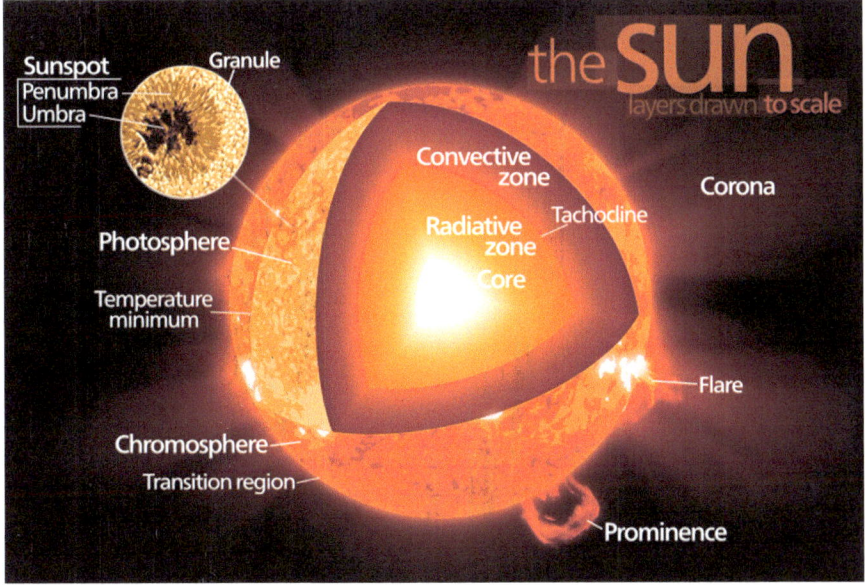

Fig. 10.1 Anatomy of the Sun (Image credit: Wikipedia)

System, such as gold and uranium, relative to the abundances of these elements in so-called Population II (heavy-element-poor) stars. These elements could most plausibly have been produced by endothermic nuclear reactions during a supernova explosion, or by transmutation through neutron absorption within a massive second-generation star (encompassing the so-called r- and s-processes).

The core of the Sun is considered to extend from the center to about 20–25 % of the solar radius. It has a density of up to 150 g/cm³ (about 150 times the density of water) and a temperature of close to 15.7 million K. By contrast, the Sun's surface temperature is approximately 5,800 K. Recent analysis of SOHO mission data favors a faster rotation rate in the core than in the rest of the radiative zone (see Fig. 10.1). Through most of the Sun's life, energy is produced by nuclear fusion through a series of steps that convert hydrogen into helium.

The core is the only region in the Sun that produces an appreciable amount of thermal energy through fusion. Indeed, a whopping 99 % of its power is generated within 24 % of its radius. When we move out to 30 % of the radius, fusion has petered out almost entirely. The rest of the star is heated by energy that is transferred outward by radiation from the core to the convective layers just outside. The energy produced by fusion in the core must then travel through many successive layers to the solar photosphere before it escapes into space as sunlight.

Examining the intensely brilliant Sun safely can be done in a number of ways, either by solar projection or by using a full aperture solar filter.

As we said earlier, *Never observe the sun directly through any optical aid as doing so will result in permanent eye damage and/or blindness.*

Solar projection is achieved by aiming the telescope at the Sun and adjusting the telescope's position until a perfectly circular image of the Sun's photosphere is sharply focused on a white card. The card is typically positioned about a foot behind the eyepiece to catch the solar image. Using an eyepiece yielding a power of 30× or 40×, focus the projected image until it is as sharp as possible.

This technique is ideal for group observing and is particularly effective if the white card is mounted on an easel or some such and shaded from direct sunlight to increase contrast. This method is particularly useful in making full-disk drawings of sunspot groups, as one only has to mark their positions on a pre-prepared card. For projection, the telescope's aperture should be restricted to about 80 mm to avoid damage to the eyepiece. Thus, small refractors are ideal for such a task. Let's take a closer look at some of the equipment amateur astronomer have at their disposal to study the Sun, the daytime star.

The Herschel wedge or Herschel prism is an optical device used in solar observation to refract most of the light out of the optical path, allowing safe visual observation. It was first proposed and used by astronomer John Herschel in the 1830s. The prism in a Herschel wedge is a trapezoidal cross section. The surface of the prism facing the light acts the same as a standard diagonal mirror, reflecting a small portion of the incoming light at 90° into the eyepiece.

The trapezium-shaped prism at the heart of the wedge refracts the remainder of the light gathered by the telescope's objective away at an angle, reflecting about

Fig. 10.2 A Baader Herschel wedge used for white light solar viewing (Image © Darryl Kirby. Used with permission)

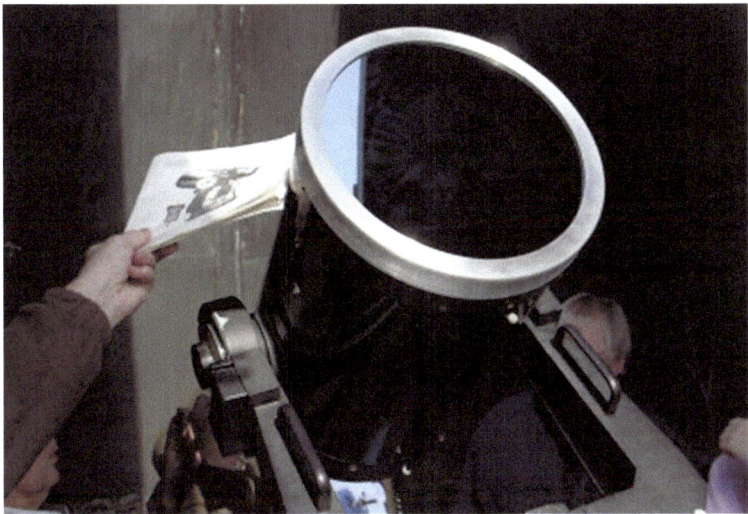

Fig. 10.3 Thousand Oaks glass solar filter (Image © Steve Wainwright. Used with permission)

4.6 % of the light that passes through one of the prism faces that is flat to 1/10 of a wave. Thus, about 95.4 % of the Sun's light and heat goes into the prism and exits through the other face and out the back door of the housing. Although the prism decreases the intensity of the light, it does not affect the visible spectra, resulting in a more accurate spectral profile that can be filtered to bring out certain details. It is an excellent, though more expensive, alternative to using white light filters, which, despite their name, inherently must block certain visible spectra.

A proper solar filter, specifically a full aperture solar filter, is adjusted to fit snugly over the front of your telescope, where it safely reduces the intensity of sunlight before it enters the tube. The most durable pre-filters are manufactured from optical plane-parallel glass coated with nickel chromium alloy. Companies such as Thousand Oaks are one of the leading manufacturers of these filters and can be purchased in a variety of sizes. Prices range from about $50 for the smallest units up to $200 for the largest apertures (~200 mm).

In recent years, manufacturers have developed good alternatives to glass-based solar pre-filters. For example, metal-coated thin Mylar film has largely replaced the heavier glass filters in many amateurs' kits. By far the most successful is Baader Astrosolar material, which may be bought as a bare sheet that can be fashioned into a homemade filter or a pre-mounted filter with the Baader film built in. User reports attest that the latter gives crisper images of the solar photosphere than glass-based units, especially in high power, high resolution applications.

All the full aperture solar filters discussed so far display the Sun in white light by reducing the amount of light from it across all wavelengths. There is, however, a new a suite of solar filters that block off all light except a single or narrow range of wavelengths, usually centered around the wavelength of hydrogen alpha light (656 nm).

Fig. 10.4 A simple white light solar filter made from Baader astrosolar material is an indispensible tool for Sun watching (Image by the author)

Arguably the best-selling solar 'scope of all time is the Coronado Personal Solar Telescope (PST). Retailing for about $500, the telescope has an aperture of 40 mm but is small and ultra portable, affixing easily to a sturdy photographic tripod or lightweight alt-azimuth mount. The PST can be used to see a wealth of solar phenomena including:

- *Prominences*: projections of H-alpha emissions seen off the limb of the solar disk.
- *Filaments*: prominences seen against the face of the Sun.
- *Active regions*: localized transient volume of solar atmosphere where plages, sunspots and flares may be observed.
- *Plage*: a patchy H-alpha brightening on the solar disk.
- *Sunspots*: small, dark, "cooler" areas that appear, grow and dissipate. We'll shortly take a much closer look at these.
- *Flares*: sudden eruptions in the solar atmosphere lasting minutes to hours.

H-alpha filters add a whole new dimension to the pastime of solar observing, but they are expensive, running between about $800 and $10,000, depending on the aperture required, as well as the bandwidth employed. The most common sizes have apertures in the range between 40 and 90 mm and work best on refractors. Companies like Coronado, Lunt and Daystar are the principal market leaders. Because seeing is poorer during the day than it is at night (owing to solar heating

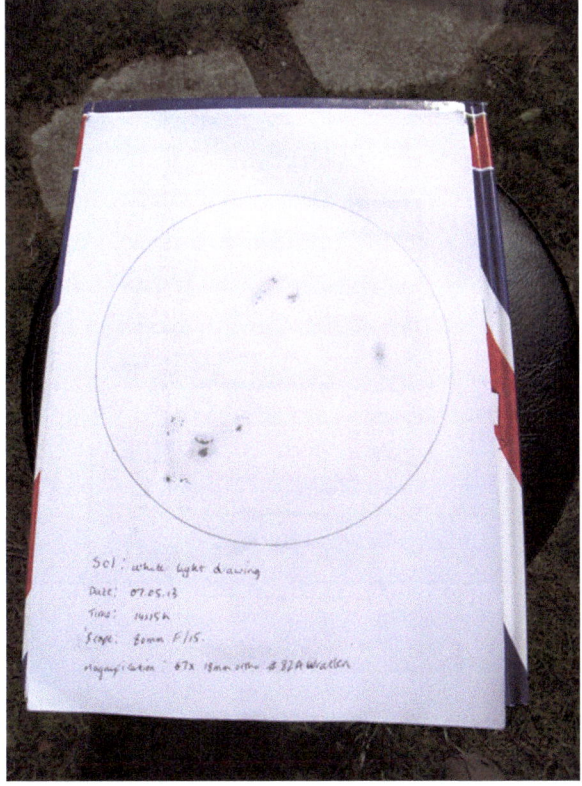

Fig. 10.5 Sketch of the Sun using a full-aperture white light solar filter (Image and sketch by the author)

of the ground and air above it), filter apertures exceeding 90 mm yield rapidly diminishing returns. In addition, it is important to note that H-alpha filters are more fragile than typical astronomy accessories, which means extra care must be taken in transporting and storing them when not in use.

Sunspots

Sunspots arise where magnetic lines of force burst through the photosphere. The magnetic field itself is generated from the moving plasma of charged particles from deep within the Sun's interior. The presence of a strong magnetic field blocks the outward flow of heat from inside the Sun, producing the cooler spot. If you look at a sunspot at moderately high power, you'll see it has a dark center, called the umbra, with a temperature of 4,000°, surrounded by a brighter penumbra radiating

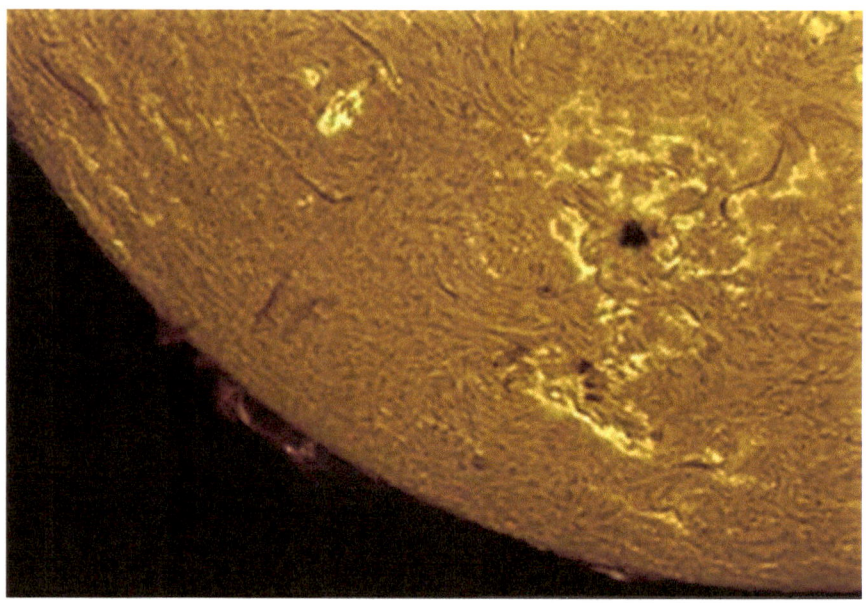

Fig. 10.6 H-alpha image of the Sun using a Coronado Solar 'scope (Image © Meade Instruments. Used with permission)

at a temperature of about 5,000°. And although appearing dark to the telescopic eye, the umbral temperature is still hotter than the surface of a red giant star, so these dark spots would actually glow a deep red if observed in isolation.

Sunspots range in size from small pores, little bigger than the granules, to enormous complex structures, many times larger than Earth. The largest spots are actually visible to the naked eye when the Sun is dimmed by the atmosphere, shortly after sunrise or before sunset, or through special hand-held solar filters.

A major sunspot takes about a week to develop to full size, then slowly fades away over a 2 week period. Sometimes, the largest sunspot groups can survive two full solar rotations. So how long does it take the Sun to complete one revolution on its axis? The answer actually depends on what solar latitude you are observing from. Because the Sun is not a solid body, it does not rotate at the same rate at all latitudes. At the equator it spins round in as little as 25 days, which slows to 28 days at latitude 45° and falling off to 34 days at the poles.

The number of sunspots on view, together with other forms of solar activity, waxes and wanes over a period of about 11 years on average, but can range from as short as 8 years and as long as 16 years. During the time of solar minimum, days or weeks can go by without seeing a single sunspot, whereas around the time of solar maximum, over a 100 spots may grace the solar photosphere at any one time. Other observers have noted that the first spots of each cycle appear at solar latitudes of about 30–35° north and south of the Sun's equator, and as the cycle progresses, they appear to form progressively closer to the equator, where the cycle begins anew.

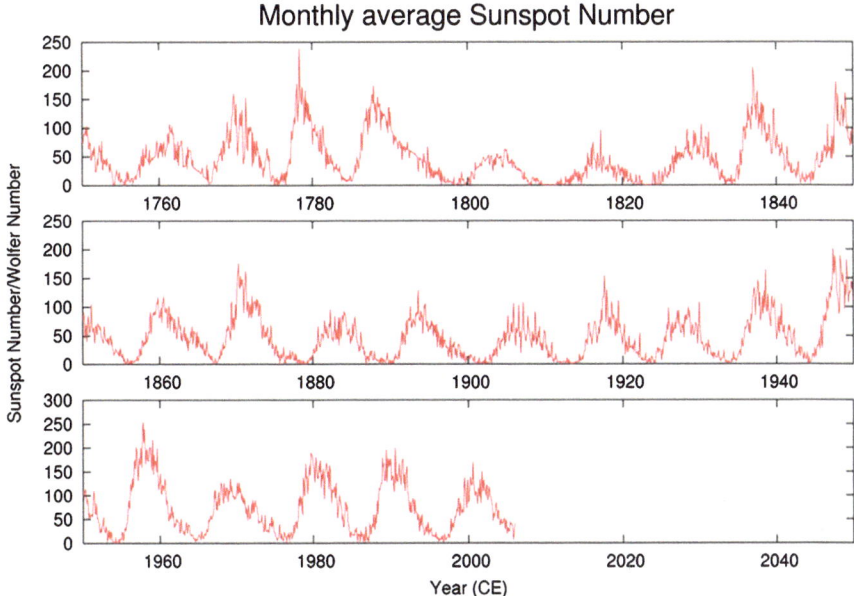

Activity: Perform a Daily Sunspot Co

After affixing your telescope with a good white light solar filter, remove the finder 'scope to prevent any accidental damage and aim your grab 'n' go telescope at the Sun. Choose a low power eyepiece at first and count the number of sunspots visible on the solar disk. A light blue 82A filter increases the visibility of the fainter spots. Of course, one may also record spots by projection, as described earlier. Try to make a daily count of the solar disk and soon you'll have built up an archive of interesting solar data.

Activity: The Sun and Climate

Find out as much as you can about how the sunspot cycle affects Earth's weather and long-term climate.

That brings us to the end of our brief overview of the Sun and how to observe it. As an accessible daylight object, our daytime star is uniquely situated for regular study with your grab 'n' go telescope. Make it part of your regular observing routine.

Chapter 11

Our Glorious Moon

About 100 million years after the formation of Earth, a Mars-sized object, Theia, struck our world, not head on but at an angle, delivering large quantities of radioactive elements to enrich Earth's core and ejecting a sizable mass of our planet's mantle material into orbit around it. The primordial Earth also lost most of its water in this momentous event (it probably had a global ocean some 100 km deep). Over several million years, the ejected mass coalesced to form the body we now call the Moon.

At first, the neonatal Luna was dangerously close to Earth, generating tides of gargantuan dimensions, but slowly (and non linearly) it inched farther and farther away from Earth, braking its period of revolution from just a few hours to the more gentle 24-h period we observe today. As a result, the Moon got tidally locked—or very nearly so—causing it to rotate once per orbit and thus keeping much the same face to us in the sky.

We have a lot to thank the Moon for. The delivery of extra radioactive elements proved vital for sustaining plate tectonic activity over billions of years. The impact event that formed our natural satellite removed a sizable quantity of water from our planet, essential for the eventual emergence of continental landmasses and thus a means of water and mineral recycling. The proximity of the Moon to Earth curbed its tendency to wobble over time, stabilizing its tilt relative to the plane of its orbit. The greater the planet's tilt, the more extreme the seasonal changes in temperature between winter and summer. This, in turn, would have put severe constraints on animal bodies, precluding the emergence of delicate creatures such as ourselves.

The Moon goes through phases, as its different sides take turns being directly illuminated by the Sun. The whole drama, from new Moon to full, takes 29 days and is thus the origin of the word 'month.' First it starts almost invisible, at new Moon, and as the days go by, the area illuminated by the Sun increases or waxes

N. English, *Grab 'n' Go Astronomy*, The Patrick Moore Practical Astronomy Series,   169
DOI 10.1007/978-1-4939-0826-4_11, © Springer Science+Business Media New York 2014

Fig. 11.1 The Moon—the other part of our double planet (Image by the author)

from the thinnest crescent, through first quarter—when it is half illuminated—through gibbous and finally to a bright full Moon. Then it begins to shrink or wane again, but in reverse. The different objects we shall visit during our lunar study are best visited at different times when the Sun illuminates them at the right angles to stand out in the telescope.

You can see a great deal with a small telescope such as the SV 80/9D. For the present study, this author will be making use of a variety of eyepieces, from 19× up to 189×. A good Moon map will also be employed to identify the various lunar features.

Let us first discuss the various topographic features visible in a small telescope.

Maria (Seas)

These are very large, dark regions that can clearly be seen from Earth with the naked eye. Consisting of very flat lunar plains, they are mostly devoid of impact craters and mountains. Even the smallest maria are a few 100 km in size and the largest extend for over 1,000 km.

Craters

These are small, circular depressions with sharply raised perimeters, ranging in size from about 5–60 km in diameter. Smaller features of this kind are referred to as craterlets. Most craters were formed long ago in the aftermath of an impact with a meteorite. A few may have volcanic origins, though.

Mountain Ranges

The lunar mountains usually occur as part of more extensive ranges. The peaks soar high above the plains. Some of the highest, like the Apennine range, are more than 9,000 m high. As the Moon goes through its various phases, the summits of these mountain ranges become dramatically visible.

Valleys, Rilles and Clefts

These consist of large chasms in the lunar surface, sometimes extending over several 100 km in extent and may be ten or more kilometre wide. Rilles appear as cracks in the lunar surface and can be excellent tests of the seeing conditions at your sight as well as the quality of your telescope's optics.

Activity: Studying a Crater

Go out one evening when the Moon shows a thin crescent. Point your telescope at the Moon and select a crater that is prominently seen. Make a drawing of its appearance at medium and high power. Follow this up by conducting further observations of the same crater as the moon proceeds through its various phases. Make notes on its changing visibility as the angle between it and the sun changes.

Observing the Moon can be a fascinating and rewarding undertaking for the amateur astronomer, even if he or she is only equipped with binoculars or with a modest telescope. That can be attributed to its great size, brightness and dynamism. In fact, the Moon presents a large apparent diameter in the sky (almost half a degree, or 30 arc minutes), is very bright, and shows profound and predictable changes in its appearance during a 29.5 day cycle, technically called a lunation.

Throughout the lunation, the Moon will practically show us the same hemisphere, the so called near-side. However, the various lunar features present therein, namely mountain ranges and craters, will always look different due to high variability of the light-shadow patterns that may occur at the Moon's surface. Even with a small telescope, the three dimensional perception of the relief of the various lunar

Fig. 11.2 First quarter Moon (Image © Mike Pearson. Used with permission)

formations will be magnificent. Occasionally, due to its orbital motion, the Moon will pass in front of a bright star or a planet. These phenomena are called *occultations*, and, as you might have guessed, can occur quite frequently if dimmer stars are considered.

However, even if the Moon does not occult bright stars or planets, fascinating celestial sights can occur if other bodies are all clustered in a restricted area of the sky. Such conjunctions are not to be missed, and are the perfect target for a first-time astro-imager. From time to time, the Moon may also be an active partner in the production of spectacular events such as eclipses, namely lunar and solar eclipses, discussed back in Chap. 9.

For those who enjoy studying astronomical history, the Moon is also the perfect object. Even a casual glance at a lunar map will show you places named after many celebrated historical figures, such as Plato and Aristotle, or Archimedes and Eratosthenes. And there are many names from nearer our time, including Nicholas Copernicus, Tycho Brahe, Johannes Kepler and Galileo Galilei—great minds now immortalized as craters on our natural satellite. And the best thing of all is that they can be easily located and studied with a small telescope.

No matter what phase it presents to us, the Moon already displays dark and bright markings—so called albedo features—when gazed upon with the naked eye. The darker regions were interpreted by earlier astronomers as seas (maria in Latin), while the brighter regions of the lunar regolith were thought to correspond to continents (terrae). Nowadays, of course, we know there are no oceans on the Moon, and the maria are just vast lava plains, but the old terminology is still being used.

Fig. 11.3 Waxing crescent Moon (Image by the author)

Except when at new or full Moon, our satellite always shows a distinct boundary between illuminated and non-illuminated hemispheres. This interface is called the *terminator.* The telescopic observation at or near the terminator is extremely informative because lunar features situated at those loci cast long shadows that bring up its relief.

Increasing the magnification provided by the telescope will allow us to view more and more details, namely isolated mountains (montes) or mountain ranges, valleys and smaller craters, and to confirm that regions occupied by maria do exhibit a lower density of visible craters when compared to terra-dominated sectors of the Moon, such as the southern highlands.

Activity: Studying the Moon Through Its Phases

Crescent Phase

About 3 days after a new Moon, a beautiful thin crescent is seen near the setting Sun. The vast oval Mare Crisium is located near the bright lunar limb. Cranking up the power of your telescope can you observe the craters Swift, Peirce and further south, Picard? The Russian spacecraft *Luna 15* and *Luna 24* landed in the southern part of the Mare Crisium.

Moving south from Crisium, nearly at the lunar equator, you'll see Mare Fecunditatis. The low Sun angle in these early days reveals a vast array of rilles and rinkle ridges (known also as dorsa). Your telescope should also show you two interesting craters Messier and Messier A. Can you see that Messier is not circular? On the south, limbward shore of Fecunditatis you should see the large (10-km) crater Langrenus. It has a central peak that should be easily visible. South of Langrenus lies an even larger crater, Petavius.

Find out how high the central peaks of Petavius rise above the surrounding lunar plains.

On Night 4 after a new Moon, the magnificent craters Atlas and Hercules begin to appear near the terminator far to the north. Hercules is particularly fascinating, as within it you can locate a smaller crater called Hercules G, which is itself 20 km across. Looking along the terminator reveals that Mare Tranquilitatis is just appearing.

As the crescent phase matures on Day 5, other prominent craters make their appearance. One of the most striking is Posidonius (95 km). Near its center lies Posidonius A, just 12 km in diameter and rilles that run either side of it. Just south of Posidonius lies the smaller (60-km) sized crater le Monnier. Here, in 1973, the Russian Moon rover Lunokhod 2 landed and drove more than 36 km across the lunar surface.

South of Mare Tranquilitatis lies Mare Nectaris and on its eastern flanks the impressions of three great craters: Theophilus, Cyrillus and Catharina. Once the shadows of the rims leave the crater floors, these craters, examined at high magnification, reveal a wealth of interesting geological features

First Quarter

Looking eastwards of Hercules and Atlas, two prominent craters of the Moon's northern hemisphere make their appearance—Aristoteles and Eudoxus (85 and 65 km, respectively)—and just south of these lies the vast expanse of Mare Serenitatis. Near its eastern shores lies two similar-sized craters called Ritter and Sabine. Once located, examine the lunar landscape just to the west of these craters. This is the location where the *Apollo 11* astronauts set foot upon the moon in July 1969. A promontory of highland material cuts off part of the connection between Mare Serenitatis and Mare Tranquilitatis. Here you can see Plinius crater and just to its northeast Dawes. Can you see a prominent dark band—Rimae Plinius—running along the top of the craters?

Just south of Plinius you will see two well-formed craters Ross and Arago. Contrast these craters with Lamont (which lies adjacent to it), which looks 'filled in' in comparison.

What does this tell us about the relative ages of Lamont and Arago?

Consulting your Moon map, locate Sinus Asperitatis, a 200-km-wide bay located between Mare Tranquilitatis and Mare Nectaris. Its southern shore is dominated by the great crater Theophilus, which is itself 100-km wide.

Fig. 11.4 The Moon at first quarter (Image © Mike Pearson. Used with permission)

Observe the southern lunar highlands. Go out when the Moon is at first quarter or in early gibbous phase. Use your telescope to probe this vast, arid wilderness in the Moon's southern hemisphere. Why does the southern hemisphere look so battered in comparison to its northern highlands? Can you make out misshapen craters? What might cause these departures from circularity?

Days 7 and 8 are great times to study the Apennine mountain range, which runs from Mare Serenitatis some 600 km south into Mare Imbrium. Telescopically, its lofty peaks are a sight to behold, towering some 5 km above the surrounding plains. The lunar Alps separate Imbrium from Mare Frigoris. On the northern flank, this alien mountain range is decorated by the magnificent Plato. Round about this time, the sun casts long shadows along the floors of this 100-km wide impact crater. On nights of good seeing, a string of craterlets can be resolved on the floor, serving as excellent tests of visual acuity and the quality of your optics.

On a night of excellent seeing, turn your telescope to the crater Plato, crank up the power and see if you can detect the small craterlets strewn across its floor. The larger members are over 2-km wide and should be fair game for apertures of 80 mm and upwards. The smallest are less than a kilometer in size and require larger apertures to disentangle.

South of the Apennines lies two small lunar seas, Mare Vaporum and Sinus Medii. Just south of Sinus Medii lies a string of three large craters: Ptolemaeus, Alphonsus and Arzachel (150, 120 and 95 km). Curiously, Ptolemaeus has no central peak, while the other two possess one.

By Day 9, the Moon is entering gibbous phase, where more than half its visible surface is illuminated by the Sun's rays. Near the center of the lunar orb lies the great crater, Copernicus (95 km), now just past sunrise. If you crank up the power in your grab 'n' go 'scope, you should be able to see that a low-lying rim on the sunward side of the crater allows light to penetrate the bowl of the crater. Magical!

Day 9 is also a good time to locate and observe arguably the most famous rille on the entire lunar surface, Hadley, found by following the natural course of the Apennine mountains northwards. Use a Moon map to locate the small crater Hadley (6 km). Hadley Rille looks for all the world like a meandering river, snaking its way across the landscape. You'll need a good 4-in. (102-mm) to have any chance of seeing this feature well. By now, also, more and more of the battered and ancient southern highlands manifests itself, affording the observer more and more vistas to see.

On Day 10, sticking to the lunar southern highlands, explore the structures of its greatest impact craters: Tycho, Longomontanus, Maginus and Clavius.

Compare the appearance of Clavius' floor with that of Longomontanus. Why does the latter appear so smooth compared with the interior of Clavius? Now look at Tycho. Why does this crater appear so much fresher than any of the others?

Gibbous Moon

By Days 11 and 12, the Moon has taken on a strong gibbous aspect, and wonderful bright rays are seen emanating from Tycho. It is around this time that the interesting crater, Schiller, is best seen to the southeast of Tycho. The eye is drawn to its strange oblong shape (70 × 180 km). Planetary scientists think it is the result of two separate impacts, with their original rims having been wiped away by the lava fields created on impact. And well up north near the terminator craters Aristarchus and Herodotus are well illuminated by morning sunlight. Using high power, look for the distinctive loop forged by Shroter's Valley. Finally, locate Mare Humorum and the large crater Gassendi. This huge (110-km) structure shows the effect of volcanic uplift and subsequent collapse, as the lava flows that once flooded it became fractured over geologic time.

Approaching Full Moon

During the full Moon, we are not seeing any shadows cast by the lunar landscape, but rather what becomes visible are sharp contrasts between light and dark regions.

Take a good eyepiece with a clean eye lens and observe the full Moon. Note its many shades of gray and brilliant white. Experiment with color filters to bring

Fig. 11.5 The gibbous Moon. South at top (Image by the author)

out subtle differences in lunar shading. Observe the bright ray craters Tycho, Copernicus and smaller but nonetheless brilliant examples like Aristachus and Kepler.

Activity: The Moon has long been thought to influence human behavior. Investigate to what extent this is true.

Transient Lunar Phenomena

Some amateurs continually monitor the Moon for so called transient lunar phenomena (TLP), which, in the main part, consist of sudden flashes of light, color changes and a mysterious obscuration of the underlying lunar features. TLP have two conceivable origins, either from within the Moon itself or the immediate aftermath of a meteoric impact event. This is an activity that requires regular and diligent observation of as much as the lunar surface as possible. Some would add that one requires an element of faith, as these events are likely to be very rare indeed. That said, the *Apollo 14* astronauts discovered gaseous emissions emanating from beneath the lunar crust.

Many TLP hunters concentrate their observations on reported 'hot spots' such as the area around the great crater Aristarchus, especially the dark canyon known as Schroter's Valley.

Grazing Occultations

Because the Moon subtends a relatively large angular size on the sky, it often occults the light of background stars. A full occultation occurs when the star totally disappears behind the Moon, whereas a grazing occultation occurs when the star dips in and out of visibility as it 'skims' over the surface of the Moon. The winking effect of such grazing occultations is caused by the star's obscuration by lunar mountains and crater rims. Keep an eye out in your monthly astronomy magazine for notices of these events.

Observing Libration Effects

Because the Moon is not completely tidally locked with Earth, the percentage of the surface area that be readily observed is greater than 50 %. In addition, the Moon's orbit is tilted by about 6.5° from the plane of Earth's equator allowing us to have a quick glimpse 'over' and 'under' the poles at the extremes of its orbit. All in all, we can see an impressive 60 % of its entire surface. Such events are called *librations*, a time when otherwise hidden parts of the lunar surface pop into view.

A really good libration target occurs near Grimaldi, a dark, mare-filled region further east than Oceanus Procellarum. If Grimaldi is easy to see, push your 'scope to the extreme lunar edge, moving south from Grimaldi Basin. You'll pick up three very delicate lines running parallel to the limb. These are actually part of the Orientele Basin, itself located on the lunar far side!

Although many amateur astronomers merely think of the Moon as a form of light pollution, the reality still is that it can be a place of endless fascination and adventure. Indeed, one could spend one's entire career staring at the world next door—the indispensible Moon.

Chapter 12

Mars, Jupiter
and Saturn

Mars, the Avenger

Of all the planets that go around the Sun, Mars has perhaps the greatest potential to captivate the human imagination. At its closest approach, it shines more brightly than any other object with the exception of Venus, its ruddy color long associated with bloodlust and war. Once a promising world for life, its small mass and low surface area (roughly the same size as the area of all the landmasses on Earth) sealed its fate early. It lost much of its water and atmosphere. Mars today is a cold, lifeless (and probably dead) desert.

Successive oppositions of Mars occur every 780 days, where its globe swells in the telescope. But not all oppositions are equally favorable for viewing. That's largely because of the eccentricity in both the Martian and terrestrial orbits, which causes the disk of the Red Planet to vary between 13 and 26 arc seconds. Only at opposition does the Martian disk appear 'full.' At other times it can exhibit a marked gibbous appearance, the visible disc illumination being decreased to just 84 % of its full value.

It was the Dutch astronomer Christian Huygens who recorded the first permanent feature of the planet Mars. "In 1856, I have seen," he wrote, "a very broad dark area in the middle of the planet Mars." We now know this triangular-shaped dark marking to be the Syrtis Major. Later, the same man recorded the planet's polar caps. It can rightly be said that our fascination with the Red Planet reached new heights of fanaticism in the late nineteenth century and early twentieth century, when some of the finest observers in the world, equipped with a new breed of high performance refractors, began to map the planet's surface in unprecedented detail.

N. English, *Grab 'n' Go Astronomy*, The Patrick Moore Practical Astronomy Series, 179
DOI 10.1007/978-1-4939-0826-4_12, © Springer Science+Business Media New York 2014

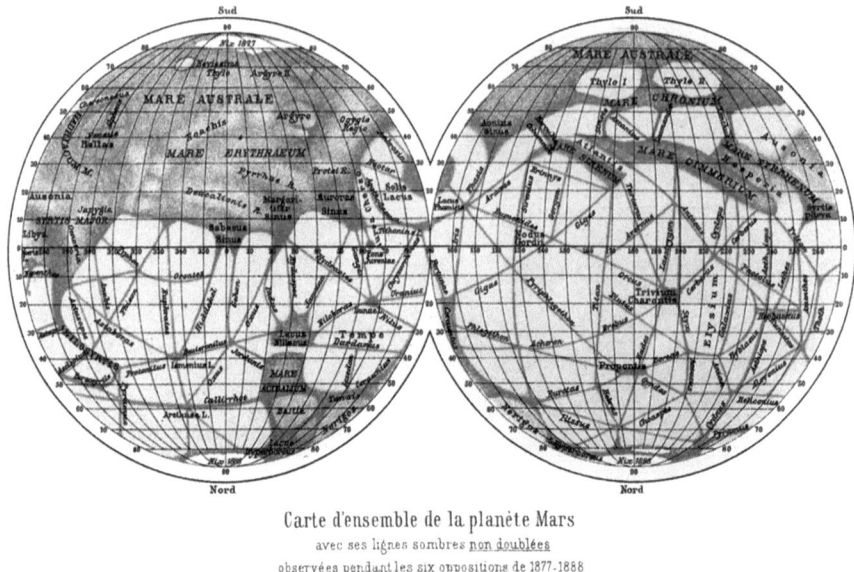

Carte d'ensemble de la planète Mars
avec ses lignes sombres non doublées
observées pendant les six oppositions de 1877-1888
par J.V. Schiaparelli.

Fig. 12.1 The Martian *canali*, as drawn by G. V. Schiaparelli (http://en.wikipedia.org/wiki/
File:Karte_Mars_Schiaparelli_MKL1888.png)

The first maps of the Martian regolith were made by very modest telescopes. For
example the German astronomers Wilhelm Beer and Johann Madler, using a
Fraunhofer refractor of just 3.75 in. in aperture, produced one as early as 1840. The
Italian astronomer, Giovanni Victorio Schiaparelli, observing the planet with a fine
8.75-in. Mertz refractor in 1879, recorded what he thought were long straight lines,
which he referred to as *canali*, which literally translates from the Italian as
'channels.'

It has been suggested that before Schiaparelli took up astronomy as a profession,
he was a hydraulic engineer and that his natural familiarity with canal works may
have prompted him to suggest such bizarre surface features. And while many of his
contemporaries saw no such canali—E. E. Barnard and Eugene Antoniadi among
them—other astronomers of note said they did.

It is arguably the observations of the wealthy Bostonian oligarch, Percival
Lowell (1855–1916), that brought Mars to worldwide attention. Observing from his
lavishly equipped observatory at Flagstaff, Arizona, Lowell used a fine 24-in. Clark
refractor to conduct studies of Mars through successive oppositions. Examining his
maps today, we see the Mars of Lowell to be crisscrossed with elaborate canals
extending from the water-rich polar caps to the arid deserts near the planet's equator. Lowell firmly believed that Mars was inhabited by a race of intelligent beings,
perhaps more advanced than our own civilization. In 1906, his sensational claims

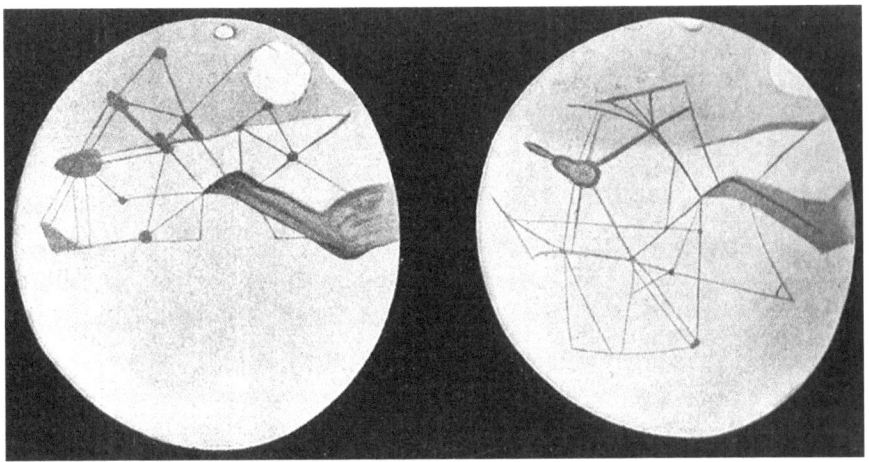

Fig. 12.2 The Lowellian canals (http://en.wikipedia.org/wiki/File:Lowell_Mars_channels.jpg)

appeared in a best-selling book, *Mars and its Canals*, followed up 2 years later with another work entitled *Mars as the Abode of Life*.

As the twentieth century moved on, more and more planetary observers became increasingly skeptical about the putative Martian canals. Lowell himself endured great ridicule in his lifetime, but the reality was, right up until the advent of the Space Age, many planetary scientists were still willing to consider simple forms of life on Mars, lichens and the like, hardy organisms that are seen to flourish in the most arid climates on Earth.

Mars observers were also curious as to what was the so-called waves of darkeni—a gradual darkening of the surface extending first near the Martian poles and extending their way towards the equator. To some, these represented the growth of vegetation as the long Martian winter gave way to spring. We now know that these are caused by ultra-fine dust that gets whipped up as the planet warms. Such dust storms can sometimes completely shroud the planet, making any surface details completely invisible.

Telescopically, Mars can often appear disappointing to the novice astronomer on account of its angular size. But when the planet approaches opposition every 2 years or so, amateur astronomers with ordinary backyard telescopes can show an abundance of detail on its surface and atmosphere. A small telescope, such as Stellarvue 80/9D, is large enough to reveal the salient features of the planet—its polar caps and some of its darker markings. Larger telescopes, of course, will show more under good conditions.

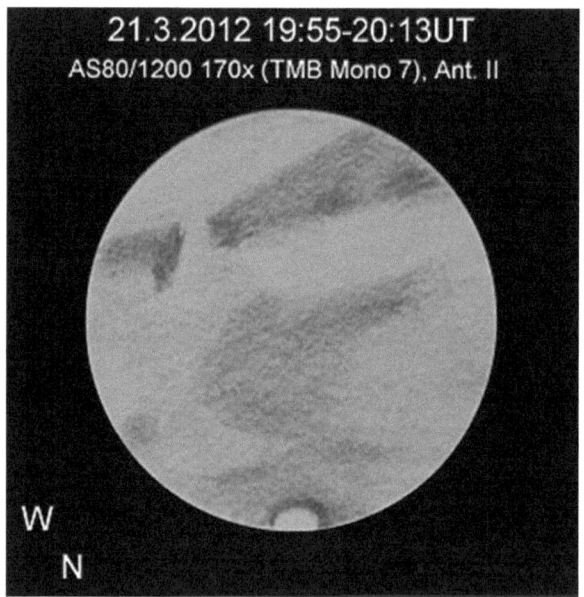

Fig. 12.3 Mars drawing (Image © Alexander Kupco. Used with permission)

Observing Mars

This planet is one that most definitely benefits from filtration. In general, choose a filter with a color *opposite* to the color of the Martian feature you wish to observe. An orange filter works wonders emphasizing the darker features, while a blue filter is excellent for observing atmospheric phenomena such as cirrus clouds on the morning limb of the planet.

Use high magnifications and good optics to get the most out of observing the Red Planet. Enlargements of between 150× and 180× are a good choice for a 80-mm refractor. Larger telescopes will allow still higher powers to be employed.

A disk drawing is always desirable, as there are usually enough surface features to make the activity worthwhile. Following are a few of Mars's more prominent telescopic features. Many more can be found using a good Martian map. See below for details.

Polar Caps: Polar caps are easily the most obvious feature to be seen on the Martian disk. Intensely white and bright, they can be seen changing in size and shape as the Martian seasons progress. On the best nights, where your telescope can take high magnification, carefully examine the borders or 'collars' of the caps for dusky markings. These were once thought to be lakes of water.

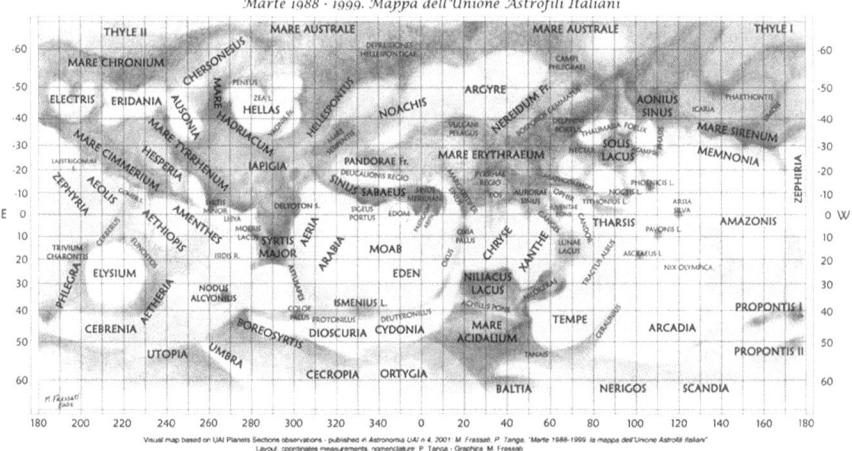

Fig. 12.4 Moon map (Image courtesy of the British Astronomical Association. Used with permission)

Syrtis Major: A large triangle-shaped structure, by far its darkest feature, usually the first marking to be picked up. Easily visible in a good 60-mm at high magnification.

Hellas: A huge impact basin on Mars, $1,800 \times 2,200$-km long and 3-km deep. It is notoriously fickle in its visibility, being almost as prominent as a polar cap at its brightest and nearly disappearing when at its faintest.

Chryse Planitia: Literally translating as the Plains of Gold, a distinctive feature, the sight of the NASA's *Viking I* lander.

Solis Planum: Formerly known as the Solis Lacus (Lake of the Sun), it is distinctly rounded in appearance.

Tharsis: The vast plain and gateway to the biggest volcanoes in the Solar System, including Olympus Mons.Study the map of the Martian surface above and pick one or two features not listed. Can you see those features on the Red Planet with your grab 'n' go telescope.

The Martian Atmosphere and Moons

Although Mars has a very thin atmosphere (predominantly carbon dioxide based), it displays a surprising amount of atmospheric phenomena, some of which are accessible to the grab 'n' go astronomer. If a blue filter is attached to the eyepiece, little of the surface is seen, since these shorter wavelengths are scattered by the Martian atmosphere. Look out for clouds (both bluish and yellow) forming on the limbs of the planet. Dust storms are also quite common on Mars and can be followed telescopically. Sometimes they became so pervasive that they shroud the entire planet in fine dust for weeks at a time!

Mars has two tiny moons, Deimos and Phobos, discovered by the American astronomer Asaph Hall using the great 26-in. Clark refractor at the U.S. Naval Observatory in August 1877. Both captured asteroids, they hurtle across the Martian sky several times each day. Though woefully beyond the reach of a humble 80-mm glass, they have been spotted in apertures as small as 10 in.

That ends our brief vigil of Mars. Be sure to turn your telescope toward it, when it shines brightly in your sky.

Jupiter, King of the Planets

Of all the worlds in our Solar System, it is arguably the gas giants, Jupiter and Saturn, that are most conducive to study. Orbiting far beyond the snowline of the Solar System, they are visible each and every year for prolonged periods of time and, when well positioned, can put on dazzling shows through the telescope.

Our civilization has much to thank these enormous worlds for. Because of their large gravitational fields, they have helped expel or deflect marauding comets and asteroids that would have otherwise have collided with Earth, causing more frequent mass extinction events. At the same time, they enjoy stable orbits far away enough from the inner planets to allow stable environments to be maintained over eons of time.

Let us begin with Jupiter, rightly regarded as King of the Planets owing to its vast size. With an equatorial diameter of 142,984 km, it is large enough to swallow 1,320 Earths whole. Because of the huge centripetal forces generated from its rapid rotation, its distance from pole to pole is 9,000 km less. This is immediately obvious to the eye, as a pronounced equatorial bulge. And, as a consequence, Jupiter is referred to as an oblate spheroid.

Spawned from primordial material from the solar nebula, Jupiter is in many ways similar to the Sun in composition (mostly hydrogen and helium). In addition, simple hydrocarbons, ammonia and phosphine, are present as minor constituents in its enormous atmosphere. Indeed, it is believed that the color of the Jovian cloud tops are derived from the interaction of these minority constituents, with sunlight creating the subtle reds, browns and blues recorded by visual observers and visiting space probes alike.

A small telescope equipped with a power of about 50× will reveal the main features of the planet—its vast belt systems and brighter zones. It will also allow you to watch the fascinating cadence of Jupiter's four large satellites—Io, Europa, Ganymede and Callisto—as they constantly change their relative position while orbiting the giant planet. Observations spaced roughly a half hour apart will easily

Fig. 12.5 The Galilean satellites of Jupiter. From left to right: Io, Europa, Ganymede and Callisto (Image courtesy of NASA)

show you changes in their relative position. You will also have ample opportunity to experience a shadow transit, where Jupiter eclipses one or more of the Galilean satellites.

Here's some data concerning these large moons:

Satellite	Orbital period (days)	Magnitude
Io	1.77	5.0
Europa	3.55	5.3
Ganymede	7.2	4.6
Callisto	16.9	5.6

Is it possible for none of the Galilean satellites to be seen? How improbable is this scenario?

Activity: The Galilean satellites are very large moons, but they have very different surface features. One aspect of their makeup that can be detected telescopically is their differing reflectivities, or albedo. Another is their differing color. By carefully studying the images of these moons at high power, can you detect subtle brightness and color differences between them. Can these be used to identify them in future observations?

Activity: Because of its large size and brightness, Earth-bound telescopists enjoy recording what they see in the corpus of a full disk drawing.

Use neither too low nor too high a magnification to get the best out of the telescope you are using. An eyepiece yielding 30× per inch of aperture seems to be the optimum. The figure below illustrates how the belts and zones on the planet are situated.

When seeing conditions are ideal, look for belts and zones, spots and ovals. Record them as best you can using a pre-prepared Jupiter blank.

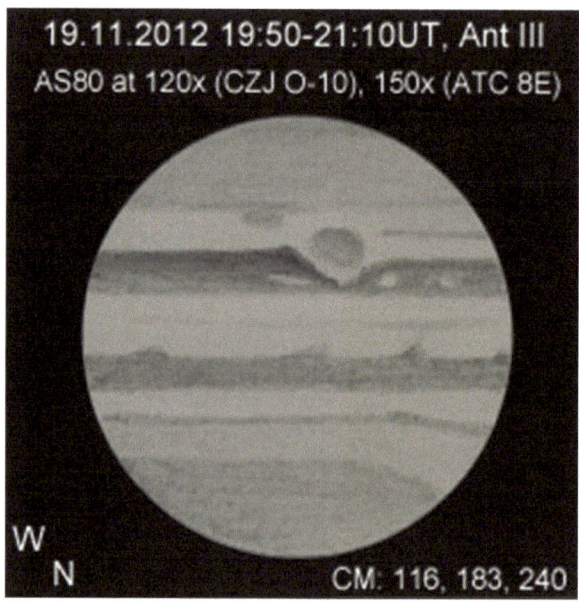

Fig. 12.6 Jupiter, as drawn by Alexander Kupco (Image © Marc McClelland. Used with permission)

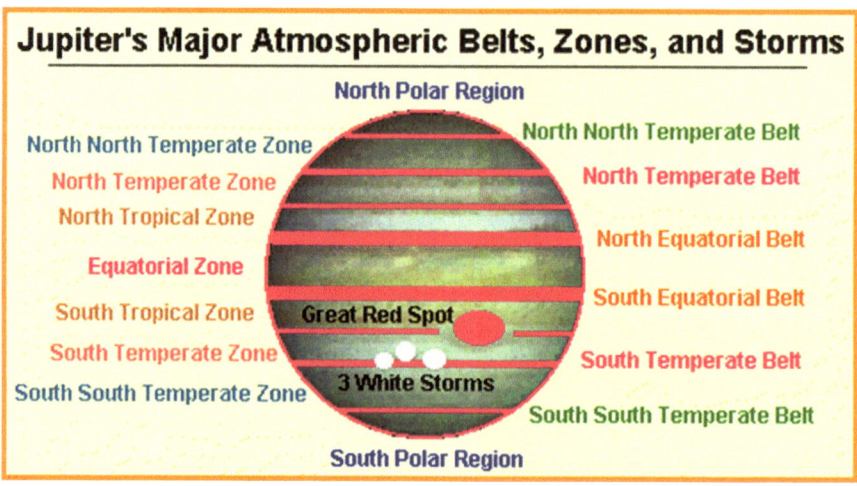

Fig. 12.7 Schematic showing Jupiter's major belts zones and the latitude of the Great Red Spot (Image credit: *Astronomy Now*)

Activity: Explore the use of color filters while observing Jupiter. Try out as many colors as possible and make records of their effects. Which filters are most effective on Jupiter?

Activity: Find out all you can about Jupiter's Great Red Spot. How long has this planetary storm raged? Has it varied in color/intensity over the years? Will it ever disappear?

Saturn, Ringed Jewel of the Solar System

Like its big brother Jupiter, Saturn is a world that has drawn the finest planetary observers over the years. Smaller and further way than Jupiter, it is very similar in composition. Yet unlike mighty Jove, it sports a beautiful system of rings (Jupiter also has a ring system but it is quite invisible in amateur 'scopes) visible across hundreds of millions of kilometers of interplanetary space. Christian Huygens was the first astronomer to clearly see the rings.

Indeed, the modern observer, equipped with a telescope as small as 2 in. (50 mm) and employing a magnification of about 50×, should be able to see Saturn's rings at their most basic. Yet, the visibility of the rings is not constant. That's because the planet's tilt with respect to our vantage varies throughout its 29-year orbit. When the inclination is at or near 0°, only the shadow of the rings on the planet's cloud tops can be made out—and with a fairly powerful telescope to boot. On the other hand, when the planet's orbital plane is tilted its maximum 29° to our own, we see the full majesty of those rings, wide open and ripe for exploration. At their most prominent, they can be seen well in a little 60 mm refractor.

Like Jupiter, Saturn's rapid rotation (10 h, 14 min at its equator) causes its equator to bulge by about 11 % more than from pole to pole. The rings, once thought to be entirely solid, were proven to exist of smaller particles by the great Scottish physicist, James Clerk Maxwell (1831–1879). We now know they are almost entirely composed of water ice. Indeed, a back of the stamp calculation shows that there is enough water in Saturn's ring to more than fill the Mediterranean Sea.

Observing and Sketching the Ringed Planet

Because they share the same basic structure, Saturn exhibits the same system of belts and zones as Jupiter. They are, however, much more difficult to discern in comparison to nearer and larger Jupiter. A 3.1-in. refractor, delivering a power

Fig. 12.8 Saturn drawing made by Alexander Kupco

of 150–180×, can show the more prominent belts on the Saturnian globe. Larger apertures will render these details considerably easier to make out. A light blue filter is very useful for bringing out atmospheric features on the Saturnian globe.

The central part of the rings, called Ring B, is the widest and easiest to see. It is separated from its outer, fainter A ring by a small gap—just 5,000 km wide— known as the *Cassini Division*. It is quite distinct in a 80-mm telescope, but it is well seen in instruments as small as a 60-mm (2.4-in.) refractor. In addition, larger instruments will enable you to see a fainter gap, called *Enke's Division*, situated on the outer part of Ring A.

Saturn has a number of bright moons that are within reach of a small backyard telescope. The largest and arguably the most interesting is Titan, a world comparable to the planet Mars in size. It is the only satellite in our Solar System known to have an appreciable atmosphere, made mostly of nitrogen and laced with simple hydrocarbons. But although rich in organic molecules, Titan's frigid surface temperatures of −180 °C means that any reactions would be extraordinarily sluggish, precluding the emergence of any kind of biological system.

The other Saturnian satellites are very much smaller than the planet-sized Titan and thus are considerably fainter. Here's a list of some of the others of interest to the grab 'n' go astronomer.

Satellite	Orbital period (days)	Mean visual magnitude
Titan	15.9	8.4
Rhea	4.5	9.7
Tethys	1.9	10.3
Dione	2.7	10.4
Enceladus	1.4	11.8
Iapetus	79.3	10.2–11.9

Mars, Jupiter and Saturn are the most rewarding of the Sun's planets to observe and can spark a lifelong interest in their constantly changing aspect. Make them a priority for observation the next time they grace your evening skies.

Chapter 13

The Inferior Planets

The innermost planets of the Solar System—Mercury and Venus—have fascinated amateur astronomers for centuries. This author vividly remembers turning his only telescope—a small 60-mm refractor—on the intensely bright white apparition low in the western sky after sundown. Using a medium power eyepiece, the image was focused to a tiny, albeit featureless white hemisphere in the twilight. Mercury took a few more years to spot, owing to its closer proximity to the Sun and lower altitude in the sky. And while a 4.5-in. reflector was then in the author's possession, the view was far from impressive: a tiny, pink, roiling 'gibbous moon' just skirting the horizon.

Yet, quite aside from their mysterious appearance in a small telescope, it's an obviously interesting question why, among all the other planetary bodies, neither Mercury or Venus possess moons. And there are more questions for an enquiring mind, such as why did Mercury have no atmosphere to speak of, while its neighbor Venus was apparently shrouded in a dense, choking blanket of air? And why do they rotate so slowly on their axes? To a curious young stargazer, these are fascinating questions to find answers for. Yet, even in the early twenty-first century, there is much about these worlds we do not fully understand.

At the end of May 2013, a team of Japanese planetary scientists employing the most detailed simulations yet of planet formation, found compelling evidence that the proximity of a planet to the Sun seals its evolutionary fate. Comparing the formation histories of Venus and Earth, the team showed that water oceans likely never appeared on Venus for any significant length of time. Indeed, the fierce ultraviolet flux from the Sun broke down any water vapor in the early Cytherean atmosphere. The resulting oxygen quickly reacted with minerals in its crust, while the lighter hydrogen escaped to space.

N. English, *Grab 'n' Go Astronomy*, The Patrick Moore Practical Astronomy Series, DOI 10.1007/978-1-4939-0826-4_13, © Springer Science+Business Media New York 2014

Fig. 13.1 A beautiful and serene-looking crescent Venus hides its hellish reality (Image ©
Mike Pearson. Used with permission)

What's more, their results showed that this early loss of water was responsible
for Venus maintaining a molten crust for a prolonged period of time compared with
Earth—as long as 100 million years as opposed to just a few million for our own
world. The differences between the planets were enough to cause the researchers to
sub-classify terrestrial planets into Type 1 and Type 2. Earth, according to the
researchers is a Type 1 planet, forming a solid crust and maintaining some (but
certainly not all) of its water inventory. Venus, in contrast, is a Type 2 terrestrial
planet, whose crust and mantle was completely parched owing to the much greater
longevity of its molten surface.

These findings have very far reaching implications for astronomers studying
exoplanets, that is, planets orbiting other stars. In particular, many of the extra-solar
planets identified to date orbit their parent stars closer in than Earth does to the Sun.
Many of these stars have a lower mass and luminosity than our own and so would
need to be considerably closer to them in order to maintain liquid water on their
surfaces. This new study puts rather severe constraints on the long-term viability of
planets orbiting their parent stars closer in than Earth does in our Solar System and
thus, by implication, considerably narrows the width of any habitable zone around
such stars.

Thank goodness for Type 1 planets!

So, let us begin anew, our exploration of the 'inferior' planets, starting with Earth's 'twin' Venus. Known to the ancients as one of the wandering stars, it would have dazzled them at maximum brightness, casting perceptible shadows on those early people. The first modern people, the Romans, were apparently unaware that the 'evening star,' Vesper (or Hersperus by the Greeks) was the same as Lucifer (Phosphorus in Greek lore), the 'morning star.' Yet it was the Greek astronomers, under Roman rule, who first realized that they were one and the same object. From the early centuries of the common era, the planet was called Venus, after the Roman goddess of love.

In 1610, shortly after the development of the telescope, the Italian scientists Galileo Galilei used a homemade instrument to observe the phases of the planet, as it revolved around the Sun. And though it did not provide the killer blow to the Earth-centered cosmology of Claudius Ptolemy, it most definitely pointed toward the heliocentric cosmology, as first enunciated by Aristarchus of Samos in the third century BC.

As telescopes increased both in number and quality, astronomers all over Europe began to scrutinize the planet. In the cool of the wee small hours of August 18, 1686, Giovanni Domenico Cassini examined Venus and spotted what he thought to be a satellite at a distance roughly 60 % of the apparent size of the planet. Cassini even reported seeing the same phase in the Cytherean moon. Other experienced observers reported similar things. James Short, employing a much improved reflecting telescope, saw the Venusian satellite on October 23, 1740. Other observers of notoriety added their names to the list of sightings of the elusive satellite of Venus, including Tobias Mayer (1723–1762) and the noted comet hunter Jacques L. Montaigne (1716–1785). Others, though, including their contemporaries Johann Schroter and Sir William Herschel, never saw a satellite. Why did some observers report it and others not? That was a mystery worth investigating, and the first stab at it came from the musings of the British astronomer Admiral William Henry Smyth (1788–1865), who suggested that the satellite was very small compared with its parent planet, but parts of its surface were much more reflective than other parts. As this reflective part came into view it was visible but not while out of view.

A raft of new sightings of the putative Venusian satellite were recorded throughout the nineteenth century, too. In 1882, Belgian astronomer Jean Charles Houzeau (1820–1888), having observed the object, proffered the idea that it was not a Cytherean satellite but a novel planet, which he named Neith, orbiting just beyond Venus. Even one of the greatest observers of all, Edward Emerson Barnard (1857–1923) claimed to have observed a seventh magnitude star not recorded on any charts on the evening of August 13, 1892, using the great 36-in. Lick refractor.

With hindsight and with the evidence from a flotilla of spacefaring robotic emissaries from the Planet Earth, any respectably sized moon of Venus would have detected by now. Ghost imaging in uncoated telescope lenses and or other optical culprits are the likely explanation for the false positive reports.

Venus's Mysterious Atmosphere

Earth-sized Venus has a massive, corrosive atmosphere made almost entirely of carbon dioxide, which created the runaway greenhouse conditions that transformed it into the hot-house world (90 bars, surface temperature 470 °C). It seems obvious to us, of course. What else could that intensely bright white cloak be? Yet, this was not always taken as granted. Up until the early twentieth century, astronomers considered it reasonable to consider the surface of Venus to be a lush Garden of Eden, with seltzer oceans, steamy jungles and even dinosaurs!

Because of their closer proximity to the Sun, both Mercury and Venus, every now and then, cross the face of the Sun, as seen from Earth. Such transits occur rarely, but with great regularity, and have afforded great opportunities to advance our knowledge of the planet Venus, as well as ours.

Such an opportunity presented itself on June 6, 1761, when an international effort was made to undertake a measure of the Earth-to-Sun distance using a technique first suggested by Sir Edmond Halley that involved a comparison of various apparent paths of the transit as measured from different terrestrial latitudes. As a leader of the Russian Academy, Mikhail Lomonosov (1711–1756), among others, spearheaded a truly global observation effort, which comprised an impressive body of 170 astronomers scattered over 117 stations around the globe. As well as providing his two cents of data, Lomonosov was eager out to find out if Venus might have an atmosphere. If one existed, he reasoned, it would bend the solar rays to produce a discernible aureole (annulus of light), occurring only at the very beginning (ingress) and very end (egress) of the transit. Remarkably, Lomonosov only used a dangerously weak filter to protect his eyes—woefully inadequate by today's standards—as he expected the aureole to be rather faint. Nonetheless, at the eyepiece of a small Dollond achromatic refractor he beheld, during brief glimpses, that beautiful arc of light contrasting the planet's shadow at the end of ingress and at the beginning of egress.

Don't try that at home!

Other astronomers, observing the same transit, reported the Venusian aureole but had not worked out its significance. The following month, Lomonosov published his findings and offered a comprehensible explanation of how the atmosphere refracts light to produce this rare and remarkable apparition from the book of nature.

"Venus," he proclaimed, "is surrounded by a distinguished air atmosphere, similar (or even possibly larger) than the one around Earth."

When similar observations were made of Mercury, they found no such aureole, suggesting it had little or no atmosphere. But was the Venusian atmosphere completely impenetrable? Was it possible to peer through the planet's clouds to make out surface markings? With new tools available to the astronomers of the nineteenth century, some believed so. Using yellow filters, observers could cut out some of the intrusive glare from the planet in the hope of seeing more subtle features. Percival Lowell, whom we met earlier in connection with the mythical Martian canals, reported seeing linear features on Venus. What his 1896 drawings seemed to illustrate were spoke-like projections dominating the western hemisphere of the planet. Other observers of note confirmed Lowell's spokes. But Antoniadi in Italy vehemently denied their existence. And that seemed to be the end of the matter until in 1951, when the British planetary observer, Richard M. Baum, not being familiar with either Lowell's or Antoniadi's work, appeared to 're-discover' them.

What's more, when planetary observers began employing blue and violet filters, these spokes were seen with greater clarity and seemed to have gained acceptance when a number of professional astronomers, using ultraviolet sensitive emulsions, detected similar features.

Observing Venus with a Small Telescope

The planet Venus, owing to its fairly large size, highly reflective atmosphere and close proximity to Earth, is a rather good target for a small backyard telescope. Arguably one of the greatest visual observers of Venus in the modern era is Richard M. Baum, from Chester, England, who produced some amazing drawings of Venus (and Mars) using an object glass of 11.5-cm aperture. Yet with his patience and attention to detail, Baum was able to record some novel features of Venus that many less experienced observers seek out today. In this capacity, an 80-mm achromatic refractor coupled to some judiciously chosen filters, can do surprisingly well at showing some of the more prominent features of the Cytheran atmosphere.

As any observer dedicated to studying Venus knows, filters are an indispensible tool in reducing glare and highlighting subtle differences of shade in its massive swirling atmosphere. If you're using a small telescope (say less than 10-cm aperture), high transmission filters such as light blue 82A and light yellow #8 are tried and trusted tools. Other observers recommend a red (#25) and blue (#44A). Venus can be viewed favorably in broad daylight, if you know exactly where to look for it. Modern go-to telescopes greatly facilitate the search. Red and yellow filters work better when Venus is observed in a blue sky.

Activity: If you're using an equatorially mounted, non-go-to telescope you can point it in daylight directly at Venus using the setting circles on your mount. Just move the telescope to the quoted celestial coordinates of the planet.

If you're using a mounted alt-azimuth instrument, you can find Venus by first observing a bright star in the night sky with the same declination as the planet. The 'scope is then pointed at this star. Take a note of the time, and be sure to clamp the telescope so that it cannot be moved. Next determine the right ascension (RA) of the same star. This will allow you to calculate the time when Venus will be seen in the telescopic field:

$$0.997T + \left(RA \text{ of the star} - RA \text{ Venus} \right)$$

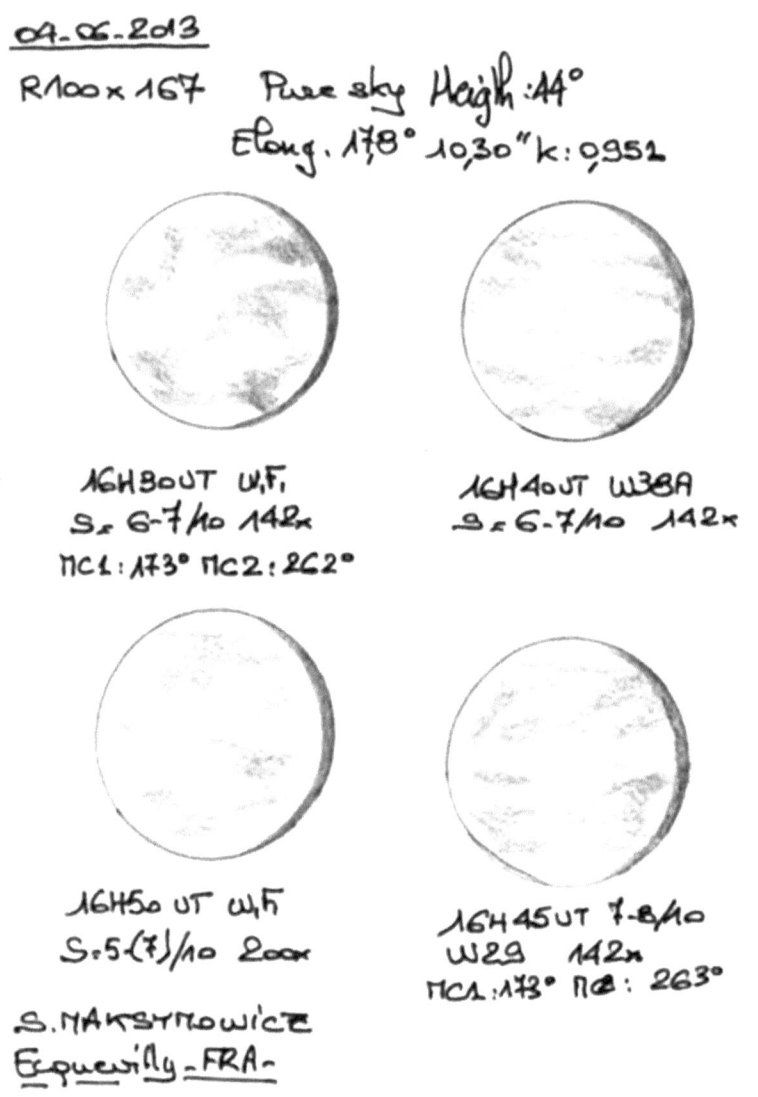

Fig. 13.2 Venus can be successfully observed in daylight (Drawing courtesy of Stanislas Maksymowicz. Used with permission)

where T represents the local mean time when you pointed the telescope at the star during the night. The factor 0.997 is to adjust for mean solar time. If RA star – RA Venus is negative, simply add 24 to make it positive again. This formula works surprisingly well so long as the interval between observing the star and Venus is less than 24 h.

French amateur astronomer Stanislas Maksymowicz managed to make some high-quality daylight drawings of the planet using a 10-cm (4-in.) Tal refractor.

Seeing any contrasted features in the Venusian atmosphere will put your observing skills to the test. To the casual eye, the planet looks pure white. But by experimenting with filters and magnification, you may begin to discern bright and dusky markings that may be classified into a variety of categories:

Banded: These are seen streaked roughly at right angles to the line joining the planetary cusps.

Radial: These are spoke-like projections that appear to radiate from the so-called sub-solar point (that imaginary spot on the planet where the Sun is perceived to be directly overhead).

Amorphous: Dusky features that have no particular shape.

Look closely at the poles of the planet. Can you make out if they are slightly brighter than the atmosphere at lower latitudes? If so, you've discovered the polar hoods. These were first seen by the German astronomer Franz von Paula Gruithuisen in December 1813, using a small refractor. The existence of the bright polar hoods is anomalous in its own right. That's because a sphere scattering light uniformly (isotropically) ought to display slightly darker poles than at the equator, so were thought to be illusory in nature. Many of the classic observers in the ages before us, including E.E. Barnard, E.M. Trouvelot, P.B. Molesworth and H. McEwen, observed them, too. But when these hoods were captured photographically in 1927, their basis in reality was affirmed.

The Ashen Light

Almost as long as people observed Venus with telescopes, they have noted an eerie grayish light emanating from the dark side of Venus. It wasn't until 1714 however that this night time glow was first accurately described by the English observer William Derham, who famously compared it to Earthshine. Though somewhat elusive and unpredictable, the so-called Ashen light is usually most easily discerned while the planet is going through inferior conjunction. Its origin is still somewhat of a mystery.

It can't be analogous to Earthshine, as Venus does not possess a satellite and Earth is just too far way to exert any effect on its atmosphere. A few years back, space probes orbiting Venus detected lightning in Venus' corrosive atmosphere. Another explanation, first put forward in the 1980s by astronomers D. A. Allen and J. Crawford, invokes the presence of sulfuric acid droplets, concentrated in a cloud deck roughly 50 km high. Being highly reflective, these droplets might reflect light from the sunlit side of the planet into the dark side, creating the eerie glow we sometimes observe through the telescope. Whatever the explanation, it's worth a serious visual study.

In the days before radar, astronomers attempted to measure the rotation rate of Venus by looking for some atmospheric features permanent enough to follow from cycle to cycle. But the ephemeral nature of the planet's cloudy upper atmosphere made any such venture an exercise in futility. Some authorities argued for a 25-day rotation period and others argued for a 24-h period, just as Earth and Mars has. Controversy raged until, to the surprise of all, radar measurements revealed that the Doppler broadening of their returned echoes implied a sluggish rotation rate— 243 days to be precise! That means that a day on Venus is longer than its year (just 222.6 days)! Intriguingly, radar studies of this kind established that the world next door rotates on its axis in a retrograde fashion—unlike all the other planets in our Solar System.

What might have caused Venus to spin in the opposite way to the other planets?

One explanation for its slow, retrograde rotation is that the planet experienced a collision with a large planetary embryo early in its history, just like Earth did in forming the Moon. The impact, apparently, had the effect of reducing the planet's spin rate to almost nothing!

How do planetary scientists define 'north' and 'south'?

If you've done your homework, you'll have discovered that these cardinal directions are formulated on the basis that the planet moves west to east. Thus, by our self-imposed definition, what we see through a small telescope as the north pole of the planet is really the south pole.

Another enduring mystery about Venus is its apparent lack of an indigenous magnetic field. A simple calculation from its mass and volume shows that its average density is practically identical to that of Earth's at 5.2 g/cm^3. Thus, in terms of bulk properties, Venus seems similar to Earth. It then ought to have a metal-rich (most likely iron) core. Yet it has an extremely feeble magnetic field. This is almost certainly the result of its painfully slow rotation, which fails to generate a proper dynamo. Lacking a magnetosphere, the planet receives no protection from the solar wind and as a result, it's continually bombarded by these subatomic particles maintaining its upper atmosphere in a permanently ionized state.

While most astronomers disregard Venus as an uninviting object to observe, knowing how mysterious this planet next door really is can only serve to enhance its allure through the eyepiece.

Chapter 14

Variable Stars, Comets and Meteors

Variable Stars

Certain stars change in brightness over time and are appropriately called variable stars. Unlike many other fields of astronomical enquiry, amateurs can still make valuable contributions to the study of these fascinating suns. The observer estimates the brightness of the variable by comparing it with nearby stars of known magnitude. The observations are then plotted on a graph to create what is known as *a light curve*, illustrating how the stars' brightness varies with time. Such data can reveal a great deal about the nature of the star under study.

Anyone can become a variable star observer, and you don't need expensive equipment to do so either. The idea is that you watch a suspect star and compare it to two other stars located either within the same field of view or as close to the candidate star as possible, one brighter and one fainter. By comparing the brightness of these comparison stars with your putative variable star, you'll be able to monitor any changes over a period of time. Becoming a dedicated variable star enthusiast takes patience—bucket loads of patience—but as a result of your efforts, you'll have gained an excellent knowledge of the night sky.

Here are the light curves of three well-known variable stars.

There are many reasons why the brightness of a star may vary. The most obvious way this can occur is through changes in the star's light output. In other cases, such as Algol in Perseus, the star is a member of a close binary system in which one star periodically eclipses the other. For this to happen, the star's orbit must be orientated edge-on to us. The Italian, Geminiano Montanari (1633–1687), was the first astronomer to point out the variability of Algol in the 1660s, but it was the young

N. English, *Grab 'n' Go Astronomy*, The Patrick Moore Practical Astronomy Series, DOI 10.1007/978-1-4939-0826-4_14, © Springer Science+Business Media New York 2014

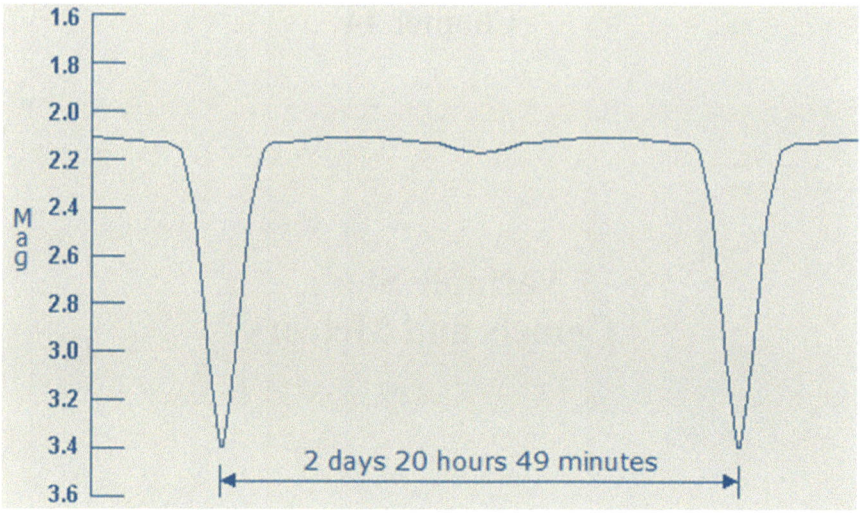

Fig. 14.1 Light curve of Algol (Image © Marc McClelland. Used with permission)

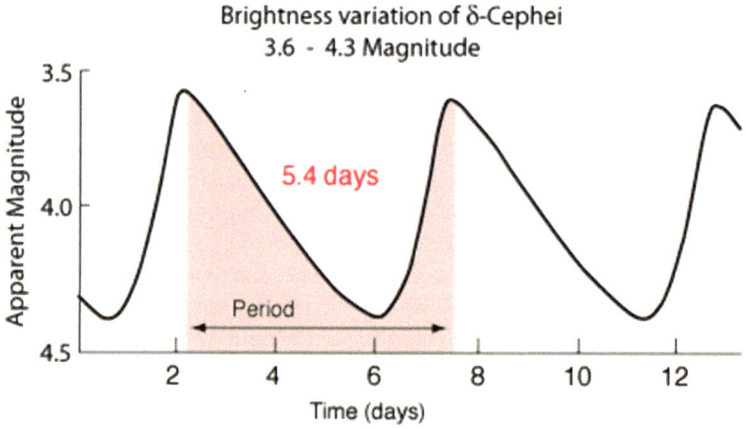

Fig. 14.2 Light curve of Delta Cephei (Image © Marc McClelland. Used with permission)

Englishman, John Goodricke (1764–1786), who first drew the world's attention to its periodic nature.

Algol consists of a blue-white main sequence star, from which most of the light is derived, orbited by a fainter, orange subgiant. Every 2.87 days Algol's brightness drops from magnitude 2.1 to 3.4 in 5 h, as the fainter star obscures its brighter counterpart and returns to normal after another 5 h as the star re-emerges.

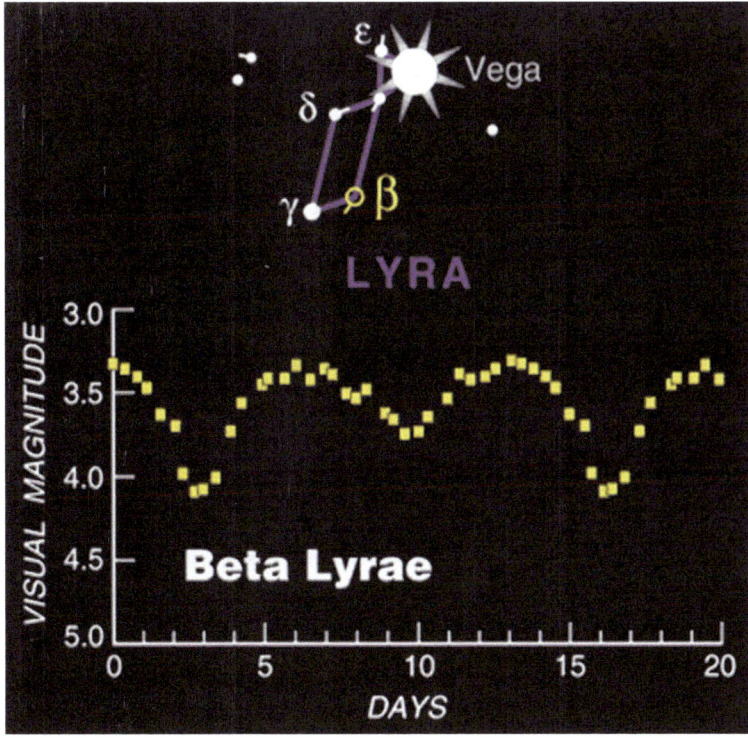

Fig. 14.3 Beta Lyrae light curve (Image © Marc McClelland. Used with permission)

Carefully examine the light curve of Algol. Did you notice a slight dip in brightness about half way through its cycle? What might cause this?

The answer is that this coincides with the brighter star passing in front of the fainter secondary. This is known as a *secondary eclipse* (as opposed to the primary). The drop in brightness is far too low for the human eye to detect, though.

Beta Lyrae is one of the easiest variable stars to find and observe, and also one of the most fascinating from a scientific point of view. Like the much better-known Algol, Beta Lyrae is another *eclipsing binary*. While Algol's eclipses are dramatic but brief, Beta Lyrae's behavior is less dramatic, varying more gently and continuously over its 12.9-day period, as shown by its light curve above. If you look carefully, you have a pretty good chance of catching this star when it's significantly below maximum brightness on any given night. Incidentally, Beta Lyrae also doubles up as a fine double star for small telescopes.

Like Algol, you can estimate Beta's brightness by comparing it with the other five stars that make up Lyra's distinctive geometric shape. At its brightest, Beta is a near-twin of nearby Gamma Lyrae. But at the deepest part of its primary eclipse, Beta is the faintest (by a small margin) of Lyra's signature stars.

Astronomers have discovered that Beta Lyrae's gentle variation results from its components being situated so close to each other that they distort each other's shape and, as a consequence, exchange material with each other.

Delta Cephei is an entirely different kind of animal, though. This star varies in luminosity because of a fundamental change in its size. Delta Cephei is the founding member of the so-called *Cepheid variables* used to measure intergalactic distances.

Cepheid variables are yellow supergiant stars that go through one cycle of pulsations in periods ranging from 2 to 40 days, varying in brightness by up to one stellar magnitude. Delta Cephei itself ranges in brightness from 3.5 at its maximum to 4.4 at its minimum and so is easily detected by the human eye from a dark sky. Good comparison stars to use with this variable include Zeta Cephei (mag +3.40) and Epsilon Cephei (+4.2).

The accepted explanation for the pulsation of Cepheids is called the Eddington valve, or κ-mechanism, where the Greek letter κ (kappa) denotes gas opacity. Helium is the gas thought to be most active in the process. Doubly ionized helium (helium whose atoms are missing two electrons) is more opaque to radiation than singly ionized helium. The more helium is heated, the more ionized it becomes. At the dimmest part of a Cepheid's cycle, the ionized gas in the outer layers of the star is opaque, and so is heated by the star's radiation, and due to the increased temperature, begins to expand. As it expands it cools, and so becomes less ionized and therefore more transparent, allowing the radiation to escape. Then the expansion stops and reverses due to the star's gravitational attraction. The process then repeats.

The importance of Cepheid variables is that their period of pulsation is directly related to their absolute magnitude; the brighter the Cepheid the longer it takes to complete one pulsation cycle. Thus, astronomers have an easy way of finding these stars' absolute magnitude, simply by observing their period of pulsation. When the absolute magnitude is then compared to the apparent magnitude, the Cepheid's distance can be easily computed. In this way, Cepheid variables are one of the best "standard candles" available to astronomers, allowing them to measure distances to faraway galaxies.

Another important class of pulsating stars are the so called *RR Lyrae variables.* These are old blue stars that often inhabit the distant globular clusters and can vary between 0.5 and 1.5 visual magnitudes. The prototype, RR Lyrae, varies between magnitudes +7.1 and +8.1 over a period of just over 12 h. Like the classical Cepheids, the RR Lyrae variables have similar absolute magnitudes, making them invaluable distance indicators.

Long Period Variables

Red giant and supergiant stars are old stars that frequently turn out to be variable. They do pulsate, but not with anything like the same regularity as the Cepheids and

RR Lyrae variables. The most abundant and well-known examples of this type of star are the so-called *Mira variables*, after its prototype, Mira, located in the constellation of Cetus. Mira exhibits a huge variation in brightness, varying between magnitude +3 and +9 over an 11-month cycle.

Activity: Locate Chi Cygni in the neck of the celestial Swan. Note its color in the telescope. This is yet another example of a Mira-type variable star. Find out by how much it varies in brilliance over its period.

What are semi-regular variables? Can you name some common examples that you can see through your telescope this evening?

Novae

As the author was preparing this section, a bright new star appeared in the constellation of Delphinus. It brightened from an obscure 17th magnitude star to a naked-eye object, its blue-white light culminating in a fourth magnitude glow. Over the days and weeks after it burst upon the stage of the sky, the star slowly faded back into the darkness from where it came. This was Nova Delphini, discovered by Japanese amateur astronomer Koichi Itagak.

Novae are arguably the most spectacular of variable stars that suddenly and quite unexpectedly brighten by up to ten stellar magnitudes or more, sometimes becoming visible to the naked eye where no star appeared before (hence their name). Astronomers have determined that they are old, faint stars that have undergone a temporary outburst. Despite their name, it's important to stress that they are not related to supernovae, which, as we've discussed previously, are massive stars that explode for different reasons.

Novae are close binary star systems, where one member has already evolved into a white dwarf. Gas spilling from the companion star onto the white dwarf becomes compressed and superheated and is eventually thrown off in a spectacular explosion. Many novae recur, that is, they undergo more than one recorded outburst, notably RS Ophichi and T Pyxidis. Indeed, given time, all novae probably recur.

A nova rises to a maximum brightness in a few days. Amateur astronomers are often the first to spot such eruptions before notifying the professional astronomers. Perhaps the best-known British nova hunter of all time, George Alcock (1912–2000), is said to have memorized the positions and brightness of over 30,000 stars, allowing him to quickly spot an eruptive nova! After a few days or weeks at maximum, the nova begins to fade back slowly to its original brightness.

Following the progress of a nova is arguably one of the most fascinating and worthwhile activities available to amateurs. What's more, nova hunting can be done with nothing more complicated than ordinary binoculars. Patience and time are the other essential ingredients.

We will now take a look at some of the more prominent variable stars on show throughout the astronomical year.

Lambda Tauri: Fluctuates between magnitude +3.3 and +4.2 over a period of 3.95 days. This is another eclipsing binary (like Algol). Use either Gamma Tauri (magnitude +3.68) and Xi Tauri (magnitude +3.74).

Eta Aquilae: A classic Cepheid variable, fluctuating between magnitudes +3.5 and +4.3 every 7.18 days. Use beta Aquilae (+3.7) as a comparison star.

Chi Cygni: This Mira-type variable brightens from +14 magnitude to a respectable +3.5 every 407 days.

R Cassiopeaiae: This long-period variable brightens from +14 to +7 over a period of 431 days.

SS Cygni: Known as a peculiar dwarf variable, brightening irregularly from magnitude +12 to +8.

R Serpentis: Located near the bright star Beta Serpentis, this long-period variable fluctuates between +5.7 and +14 every 357 days.

Y Ophiuchi: Another Cepheid, varying from +6.5 to +7.3 every 17.12 days. This is a rather long period for a classical Cepheid.

That concludes our short discussion of the variable stars. Of course, there are many more that you can visit should this brief overview manage to whet your appetite.

Comets and Meteors

Fear not the Comet,
Icy wanderer from afar.
Chock full of toxic things,
Less suited for life than for the car.
But when sunlight chases its volatiles away,
Only rubbly residue remains,
Burning up in the Sky,
Great Fires on high,
Enunciating the Shooting Star!

Fear not the Comet,
For portend it doesn't.
Just going about its way,
To and from the domineering Sun.
But once or twice in a lifetime,
If fortune smiles this way,
Our feeble predictions go queerly awry,
Promoting Itself from faint fuzzball
To an Apparition that commandeers the Sky.

Fig. 14.4 Comet Lovejoy as it appeared on Boxing Day, 2013 (Image © Denis Buczynski. Used with permission)

Fear not the Comet,
For Caesar's grace it did not demand.
The Senators did their own bidding,
in conclava; behind closed doors.
It knows nothing of Itself.
Effortlessly doing what it cannot do,
Astride the Comet,
It no longer fears its own shadow,
Yet, neither was Neptune spawned from its oceans.

Arguably one of the most amazing things anyone can witness is the sight of a bright comet in the night sky. There was great anticipation to the return of Comet Halley generated as it approached the Sun after 76 years from the depths of the outer Solar System. But comets are like cats; they may put on a show or they may not. Alas, Halley never did get that bright, but others—Hale-Bopp and Hyakutake— were more spectacular and really captured the imagination of the general public.

Comets are, for the most part, city-sized, loosely knit assemblages of ice, frozen gas and dust that round the Sun in highly elongated orbits. They are thought to represent some of the oldest and most pristine substances that condensed out the solar nebula some 5 billion years ago. Unlike the planetary bodies, the orbits of which are only slightly inclined to the plane of the ecliptic, the trajectories of comets are

Fig. 14.5 Comet 17P/Holmes as it appeared in the UK in November 2007 (Image by the author)

often highly inclined (between 0° and 90°) to the same plane. They return to the warmth of the inner Solar System at intervals ranging from just a few years to many thousands of years, slowly brightening as they approach our world and transforming themselves into glowing apparitions for a few weeks before fading away into obscurity.

Astronomers have determined that comets originate from two locales within our Solar System. The Kuiper Belt is a vast annulus that extends beyond the orbit of Neptune and which now includes the dwarf planet, Pluto. Much further out, perhaps half way to the nearest star, lies the Oort Cloud, which contains a vast herd of cometary bodies orbiting in a spherical shell.

When far from the warmth of the Sun, a comet shines only by reflecting sunlight. So, because they are small, out here they are exceedingly faint. But as they approach nearer the Sun, they warm up, and their constituent icy volatiles begin to sublimate. The Sun's radiation causes the gaseous matter to fluoresce just like the neon gas within a light bulb, and in so doing, it brightens the comet appreciably.

Gas and dust released from the warming comet generate a halo or coma extending over 100,000 km into the space surrounding it. At the heart of the coma lies the comet nucleus, the only part of the body, in fact, that can truly be said to be solid, consisting of a 'dirty snowball' of ice, dust and rock. In the largest comets the nucleus may extend over a few tens of kilometers.

The steady stream of charged particles moving outwards from the Sun's surface—the so-called *solar wind*—pushes the dusty tail outwards, away from the nucleus. But it's certainly not true to say that all comets evolve tails. On average the tail stretches across the sky for one or two angular degrees (two or four Moon diameters), but in exceptional cases, it can be much longer. Comets reach their brightest—and extol their longest tails—near perihelion. For this reason, they are usually most impressive when seen immediately after sunset or just before sunrise.

There are two parts to a comet's tail. The first consists of gas blown backwards by the solar wind from the nucleus, while the other is comprised almost entirely of dust particles that have been ejected from the comet's head. The gas tail takes on a curious bluish tinge owing to the fluorescing of its constituent gas molecules, whereas the dust tail looks yellowish because the individual dust grains scatter shorter visual wavelengths better than longer wavelengths.

Comet tails always point away from the Sun, with the gas tails mostly straight. The dusty tail, in contrast, is seen to fan out as the individual grains lag behind the comet.

How large are comet tails? The answer will surprise you. A typical comet tail can extend for a 100 million km or more. In 1997, Comet Hale-Bopp exhibited a tail that extended farther than the Earth-Sun distance. But for all its magnificence, it pays to remember that the density of the particles making up the cometary tail is more vacuous than the best laboratory vacuum!

Because of their large apparent size and diffuse appearance, comets are best observed with binoculars or a small rich-field telescope. Dedicated visual hunters of comets scan the sky near the Sun just after sunset or immediately before dawn. And while the vast majority of new comets discovered are made with fully automated telescopes, every now and then a new icy body is uncovered by chance by amateurs using ordinary-sized instruments in their backyards.

A comet's brightness can be notoriously difficult to estimate, but one effective way involves defocusing the comet and comparing it to a nearby star of known brightness. Each year's batch of comets is a mixture of known comets returning to the inner Solar System and entirely new specimens, never seen before. Dozens may be visible through an amateur telescope each year; a few can brighten enough to be visible in binoculars or the naked eye.

Over 2,500 comets have well-known orbits, and as the weeks progress, more and more are being added to that tally. The comets with the longest orbital periods—typically centuries and millennia—are thought to be derived from an unseen swarm a billion strong—the Oort Cloud—that envelopes the Solar System at its outermost edges. The gravitational tugs by stars that migrate near the Sun disturb a few comets in their orbits and send them hurtling inwards to the region of the Solar System inhabited by the planets.

The inner comet cloud—the Kuiper Belt—lies just beyond the orbit of Neptune. Most so-called period comets, which have orbital periods of a few 100 years or less and thus have been observed on more than one occasion, having their origin in this inner belt rather than the Oort Cloud.

Activity: Observing a Comet with Your Telescope

The next time a bright comet graces our skies make ready your grab 'n' go telescope. Insert a low power eyepiece delivering an enlargement of between 20× and 50×. Can you clearly make out the coma? Does the comet have a recognizable tail? If so, can you estimate its size? How fast is it moving relative to the background stars?

Dust lost from a comet disperses in the space around it, but when Earth ploughs into this dust debris during its orbit around the Sun, it burns up in our atmosphere about 100 km up in the sky. Such an exciting event is called a *meteor* or *shooting star*. Usually the entire event is over quite literally in a flash that lasts no more than a second or two. In any one night, a few of these meteors are seen streaking across the sky in random directions. These are said to be sporadic in nature.

Once in a while, Earth crosses the orbit of a comet and encounters a dense swarm of dust, giving rise to a so-called *meteor shower*, where nearly all the meteors appear to come from one spot in the sky called the *radiant*. Contrary to popular belief, though, this is not the best point to look at, since the meteors will be approaching head-on from this direction and so their tails are shortest. In fact, the longest tails will be seen some 90° from the radiant.

All meteor showers are named after the constellation in which the radiant lies. For example the Perseids have a radiant in Perseus, the Orionids have their origin in Orion and so on. The following table shows the major meteor showers of the calendar year.

Shower	Date	Meteor frequency per hour
Quadrantids	January 1–6	100
Lyrids	April 19–25	10
Eta Aquarids	May 1–10	35
Perseids	August 1–20	80
Geminids	December 7–15	100

You will note, however, that the hourly frequency of the aforementioned meteor showers is just a guide, for in some years, it can be exceeded greatly. For example, an intense storm of Leonids was witnessed over the United States back in 1966, which saw an estimated 100,000 shooting stars in a single hour!

Activity: Counting Meteors

Amateur astronomers can make valuable contributions to meteor science by measuring their frequency and their brightness. The key point is that one must be comfortably seated in a reclining chair or some such, wrapped up warm and eyes fixed in a direction near (but not precisely at) the expected radiant. One of the most popular meteor showers of the year is the August Perseids, when dark skies return to northern locations once again and the nights are mild, even warm. The best displays invariably occur when the Moon is not seen in the sky.

The next time a bright comet graces our skies remember the great journey they made to decorate the sky for you. Remember also that they delivered at least some of the water that makes our planet the habitable place it is.

Chapter 15

Summer Celebrations

Summertime in northern Europe and Canada brings mixed blessings for the avid telescopist. Although the Sun is high in the sky, making solar viewing more accessible, the nights can be dreadfully short. In the UK, for example, if you happen to live north of the Midlands, high summer is invariably attended by perpetual twilight, until darker skies return at the end of July and early August. Even so, there are still many objects that can be favorably viewed with a little preparation.

Activity: Observe Noctilucent Clouds

Way up in the mesosphere, at an altitude of about 83 km, beautiful clouds, made entirely of water-ice crystals, appear in the sky between the last week of May and the second week of August, and are only visible at latitudes between 50° and 70° north. These are the wondrous noctilucent clouds, which appear to glow brightly when the Sun is below the horizon. At latitudes above 70°, the sky is too bright to ever see them.

These clouds appear in the twilight arch, when the Sun is situated somewhere between 6° and 16° below the horizon. Their hues can be quite varied—white, blue–white and sometimes golden. Noctilucent clouds are at least superficially related to cirrus, yet the latter are located at too low an altitude for them to be seen when the Sun sets this low below the horizon.

Binoculars are a great tool for observing these magnificent summer visitors to our skies. You may be able to make out veiled or billowing structures, subtle banding whirls and patches.

N. English, *Grab 'n' Go Astronomy*, The Patrick Moore Practical Astronomy Series, DOI 10.1007/978-1-4939-0826-4_15, © Springer Science+Business Media New York 2014

Fig. 15.1 The majestic noctilucent are a familiar sight around midsummer in far northern latitudes (Image by the author)

A Sky Full of Doubles!

The heavens of midsummer are full to the brim with interesting double and multiple stars that can be resolved with a small telescope. Unlike many other celestial targets these attractive pairs are largely unaffected by the encroachment of twilight. Let's start with a few easy pairings. High overhead on summer nights lies the familiar asterism of the Plough. Locate the middle star in the Ploughshare, **Mizar**, and its naked-eye companion, **Alcor**. A telescope delivering a power of 30× or above will show you that brilliant white Mizar has a widely spaced companion some 12 arc seconds away.

Locate the small constellation of Canes Venatici, the Hunting Dogs, filling a small patch of sky between Ursa Major and Bootes. Its brightest star, Alpha Canum Venaticorum, is a famous double star for small telescopes—Cor Caroli. This is showpiece object, and an 80-mm refractor does a superb job framing the system at medium powers. The primary shines an intense white color (magnitude +2.9), while the yellowish secondary shines at a more modest +5.6. The system lies at a distance of 110 light years.

Moving east into Cygnus, we track down Beta Cygni (**Albireo**), easy to locate at the foot of the Northern Cross. This is, without doubt, one of the finest celestial gems in the entire sky. Large binoculars or, better still, a small telescope, will show you the beautiful color contrast pair, comprising an aureal primary (magnitude +3.1) and emerald secondary (magnitude +5.1). That said, the colors you see may well be different.

Activity: Find out if the two stars making up Albireo are gravitationally bound or are just a chance alignment.

While you're in Cygnus, why not sojourn once again to the wide binocular pair 61 Cygni, located about two binocular fields southeast of the bright star Deneb. When the sky is darkest after midnight, you should be able to make out the ruddy nature of both stars (orange dwarf stars). **61 Cygni A** (magnitude +5.2) shines a little more brightly than 61 Cygni B (+6.1) The German astronomer Friedrich W. Bessel made this system famous back in 1838, when he measured their enormous distance (11 light years), thus establishing the terrifying distances to even the nearer stars. For one more easy pairing, locate **Alpha Herculis** (**Rasalgethi**) at the head of the strong man, Hercules. This is yet another beautiful color contrast double comprising an ochre primary (magnitude +3.5) and blue–green secondary. The primary is a red giant some 400 times the diameter of the Sun, thus rendering it one of the largest stars known.

One more source of binary 'eye candy' lies in the diminutive constellation of Delphinus, the Dolphin. Locate the diamond tip asterism making up the head of the constellation. Next aim your grab 'n' go telescope at the easternmost star making up this distinctive shape. This is **Gamma Delphini**, a wonderful sight in a small telescope. Use moderate powers to resolve its golden primary (magnitude +4.3) and yellow–white (+5.1) secondary.

Although twilit, summer skies often bring with it steady seeing conditions, enabling the observer to tease out a few more challenging binary systems. First and foremost on your list should be the famous **Double Double** in Lyra (Epsilon 1 and 2 Lyrae), easily located about 1.7° east-northeast of the bright summer luminary Vega. A little 80-mm at 180× shows them as good as you'll see in any telescope. Four perfect Airy discs! Separated by about 2.5 arc seconds, the pairs are orientated almost at right angles to each other. The glorious system, located 160 light years away, is truly one of nature's great apparitions. Bright quadruple systems of this sort are fairly rare, and this one is quite simply a nonpareil. You'll need fairly high power (say 120× or above) to do justice to this system, though. Did you notice the differing orientations of both binary pairs?

Next move your 'scope to the second magnitude star **Delta Cygni.** This binary system has a reputation for shaming many telescopes, owing to the great magnitude difference between the components (magnitudes 2.9/6.3) separated by 2.5 s of arc. The system is readily resolved in a grab 'n' go 80-mm refractor at 150–180× under good conditions. Another challenging pair for small telescopes is **Alula Australis** (Zeta Ursae Majoris), fairly easy to track down on twilit nights, shining at a magnitude +4.2. An 80-mm refractor at 150× shows the star to be peculiarly shaped. It's more like two living cells undergoing division. A 4-in. (10-cm) 'scope should cleanly resolve the companion, which is only slightly dimmer at +4.8. Last but certainly not least, have a go splitting Pi Aquilae, located just a few degrees north of brilliant star Altair. Both stars are about the sixth magnitude and are separated by a mere 1.4 arc seconds—a classic Dawes pairing if ever there were one!

High summer brings achingly short nights to many observers' shores, and the
further north one observes the worse it becomes. From 56° north latitude, the sky
doesn't get fully dark again until August, making serious deep sky observing near
impossible. There is, however, a palliative. High in the south, Boötes, the celestial
Herdsman, under the aegis of the intensely bright orange star Arcturus, provides
ample opportunity to observe both easy and challenging binary stars within easy
reach of a small telescope. What's more, you don't need fully dark skies to
observe them.

For this you can use a 3.1-in. f/15 achromatic refractor, which is capable of
serving up razor-sharp images at high power (greater than 200×), true-color
viewing required to comfortably split the tougher pairs explored here. Of course,
larger instruments will show such pairs better, but a 3.1-in. aperture is about the
minimum.

Without further ado, we'll begin with an easy target—**Delta Boötis**, the easternmost bright star making up the familiar kite-like asterism of the constellation.
Shining at magnitude 3.5, it should be visible with the unaided eye. Though binoculars easily reveal the wide open pair, the view is much more serene at 24× in a
3-in. Your eyes will meet a yellow sun-like primary separated from a magnitude
+7.8 greenish secondary by over 100 arc seconds of dark sky. A little bit more
magnification—say 50–80×—gives the best views. One might think such a widely
separated pair ought to have no physical connection, but at a distance of only 117
light years, the pair form a gravitationally bound system, with the secondary taking
120,000 years to complete an orbit.

A more challenging system awaits you just 4° north, where you'll chance upon
another, slightly fainter naked-eye star, Mu Boötis. In the 3° field served up by a
24× eyepiece, you can easily discern a near twin system to Delta Boötis—a wide
pair consisting of a fourth magnitude 'primary' separated from a seventh magnitude
'secondary' over 100 arc seconds away. But if you take a closer look at the fainter
member at high power—the 192× does nicely in the 3-in.—you'll soon realize, if
the sky is steady, that it's a binary system in its own right, comprised of a pair of
seventh magnitude stars split by a hair (2.2 arc seconds for the record!). Curiously,
all three components are now known to form a complex triple system and are
located about the same distance away as Delta Boötis.

The next pair will surprise you. For that, we'll have to move to the far northwest
of the constellation, past the familiar kite-like asterism near the border with Ursa
Major. Here you'll find the white magnitude +4.5 Kappa Boötis. Charged with a
power of 96×, a 'scope shows a seventh magnitude yellow companion only 13 arc
seconds away. Of course the eye takes them to be a bona fide couple, but careful.
Astrometric analysis suggests otherwise. The two suns are, in fact, unconnected.
Enchanting to say the least!

Now we'll move on to a real pearl. Train your 'scope back to the lovely orange
star Arcturus and pan about 7° east, where you'll soon stumble on **Xi Boötis.**
Shining austerely white at magnitude +4.8, a power of 92× reveals a pale blue seventh magnitude ball a little over 6 arc seconds away. Under steady conditions,
higher powers can help increase the image scale and increase the apparent contrast

between the pair, which orbit each other every 150 years. To some, the disposition of the secondary looks much like the faint companion to Polaris, the Pole Star.

Finally, we end our stellar sojourns with a celestial nonpareil. To see it, look for the bright magnitude +2.5 orange star **Izar** (Epsilon Boötis), located about 8° north of Xi Boötis. Steady skies and high magnifications are required to elucidate its lovely secret—a magnitude +4.6 blue–green companion separated from its primary by just 2.9 arc seconds of sky.

Now, some little telescopes can resolve pairs as close as 1.5 arc seconds *provided* they are of fairly equal brightness. But the near sevenfold difference in brilliance between Izar and its main sequence companion renders the secondary hard to see, overwhelmed as it is by the light of its primary. Optics plays a role with this system, too. A 4-in. instrument will consistently struggle with this system, but a high-quality 60-mm refractor should just do the job under good conditions. And though you can look at Izar with all sorts of instruments, from small portable 'scopes to humongous Dobs measuring fully 2 ft across, we must say that some of the finest views of Izar can be had with a 3-in. refractor.

During a recent vacation to a tiny coastal resort on the northwest coast of Scotland, this author chanced upon some fair weather. Tucked away in a shallow inlet, the early evening winds subsided gradually to a dead calm after midnight, allowing me to take advantage of exceptional observing conditions with dark magnitude +6.5 skies. On two successive nights, I was able to rack up the power on my 'scope to 276× to get a razor-sharp separation of the system. Like a budding yeast, the pale blue ball of the secondary sat on an otherwise perfect first diffraction ring of a golden orange primary. It's at moments like this that one can more fully appreciate why the famous double star observer Otto Struve was impelled to call it *Pulcherimma!*

Entering the Shallows of the Deep Sky

Contrary to popular belief, it's possible to observe an impressive cache of deep sky objects under the darker spells of twilight. Top of any list should be the **Great Globular Cluster** in Hercules (M13), arguably one the finest of its genre in all the skies. Although some inroads can be made at resolving it in a 4-in. (10-cm) instrument, much better results can be achieved using larger instruments. Choose a good steady night and high magnification to 'pull out' the individual suns from this bauble of starlight, way out in the galactic halo. A Portable 6-in. Dobsonian does a superb job resolving stars quite a way in towards the center of the cluster.

Just a wee star hop away is **M92**, the other prominent globular cluster in Hercules. Though certainly not as magisterial as M13, it is nonetheless deserving of a long, concentrated gaze when the sky is darkest just after midnight.

The constellation of the Harp, Lyra, is very well placed in the sky during the short nights of June. Its main deep sky attraction, the celebrated **Ring Nebula** (**M57**), should always be on a mid-summer menu. This magnificent planetary

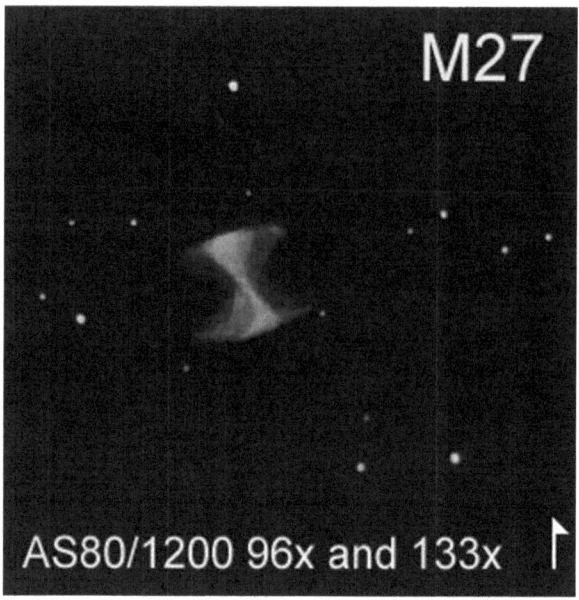

Fig. 15.2 The Dumbbell Nebula (M27) as seen through an 80-mm Zeiss refractor (Image courtesy of Alexander Kupco. Used with permission)

nebula—the incandescent glow of a long dead star—is actually quite easy to locate, roughly midway between Beta and Gamma Lyrae. An 80-mm refractor shows its elliptical shape nicely, but, like so many other objects, it benefits from greater aperture. Both narrow- and broadband light pollution filters definitely aid viewing this object. It's fun trying to evaluate the magnification at which M57 looks best. What do you find?

There's another planetary nebula well worth finding during the darker periods of twilight. It's to be found in the obscure constellation of Vulpecula, sandwiched between the more prominent summer constellations of Cygnus and Delphinus. We speak, of course, of the famous **Dumbbell Nebula** (**M27**). Despite less than perfect darkness, its ragged box shape is easy to see in an 80-mm at 60×. Larger apertures show this conspicuous planetary nebula far more compellingly, though. The view of M27 in a 9-mm 100° eyepiece (the Oculus) is quite simply stunning! As with M57, a broadband light pollution filter helps increase its contrast with the background sky.

With Ursa Major high overhead, it's always worth paying the galactic odd couple M81 and M82 a visit. This galactic odd couple is easy to find simply by drawing a line from Phad through Dubhe and extending that line about the same distance again. Using a low power ocular, you should be able to frame both these galaxies in the same field of view. The brighter of the two, **M81** (magnitude +7), is a beautiful, face-on spiral galaxy, while its fainter companion **M82** (magnitude +8) is irregular, revealing itself as an elongated blur in an 80-mm refractor. The modern high-power, hyper-wide eyepieces do a superb job at darkening the sky enough to see them quite well.

And finally, let's go after something completely different. You've just got to have a look at Herschel's famous **Garnet Star** (**Mu Cephei**), a magnitude +3.5 sun, located roughly midway between Zeta and Alpha Cephei. As its name suggests, this is one of the reddest objects visible to the naked eye, its intense sanguine color rendered blatantly obvious in an 80-mm at low power. Cranking up the power to 100× or so allows you to see this red supergiant star at its best. Its rubicund tint is caused by the presence of large amounts of carbon-rich species in its outer atmosphere, which do an excellent job at suppressing shorter visual wavelengths. Mu Cephei is a variable star, dimming from magnitude 3.4 to 5.1 over a period of 2 years.

Exploring the Glories of the Southern Skies

From most northern locations, the southern Milky Way is hopelessly subdued by twilight and the low altitude. But from lower latitudes, south of the British Isles, the glorious constellations of Sagittarius, Scorpius, Ophiuchus and Scutum rise higher in the sky, transforming the meridian into a veritable wonderland of celestial treasures. As darkness falls, the first bright luminaries appear. The bright orange star, Arcturus, is lying near the zenith and blood-red Antares, the rival of Mars shines brightly in the southeastern sky. Looking north, one's eye meets with the brilliant white supergiant star Deneb, located some 2,500 light years away, and much closer to home, refulgent Vega (25 light years) and Altair (16.7 light years) comprise the other two members of the famous **Summer Triangle.**

Center Altair in the field of view. Behold its brilliant white color. This nearby star is a rather peculiar class A dwarf star, with a radius only 60 % larger than our Sun. Although the eye can never detect it, astronomers have discovered that it revolves rapidly, taking only 10 h to complete a single revolution. Such rapid motion renders it markedly oblate.

Activity: Resolving the Companion to Antares

Antares is a one of the most magnificent stars lying in the solar neighborhood (607 light years). As its strong rubicund hue suggests, this star is a famous red supergiant with an equatorial diameter some 400 times greater than our Sun. As one might expect, such a highly evolved star is somewhat variable, fluctuating between magnitudes of 0.9–1.2 over a period spanning about 5 years. This giant star has a much fainter companion, shining at magnitude +5.4, located only 2.5 arc seconds due west of the primary. Usually described as greenish, it is a rather difficult test for a 3-in. glass. High magnification (~150×) and steady skies are necessary to prize it apart from the glare of Antares itself. The pair orbit their common center of gravity every 1,200 years or so.

Some of the other bright stars of Scorpius also have pretty companions that are easy pickings for a small grab 'n' go 'scope. First up is magnitude +2.6 **Beta Scorpii (Graffias)**, a fine optical double. A low power ocular is sufficient to resolve the blue–white 'secondary,' shining more feebly at magnitude +4.9. They look intimately connected in the telescope, but in reality the fainter member actually lies over twice as far away (1,100 light years) as the primary.

Immediately east of Graffias lies the wonderful quadruple system, Nu Scorpii. A small telescope easily resolves a northern and southern pair (A and C, respectively). A good 3-in. telescope makes light work of resolving C into two components (C and D) on a night of steady seeing. A larger telescope (say a 6-in. Dob or 4-in. refractor) reveals the final component—D—of this quadruple system. All four stars are located at roughly the same distance—450 light years by the latest measuring. Simple trigonometry shows that stars A and C are only 0.1 light years apart! C and D, unsurprisingly, are even closer—a mere ten times the mean distance of Pluto from the Sun.

It's so nice to be armed with information like this as you relax and drink up the view!

Activity: What is the smallest aperture 'scope you can employ to split the C–D components of Nu Scorpii? Use 'scopes of different aperture or make aperture masks with a larger instrument, experimenting all the while with different magnifications. What did you discover?

Move your telescope south into the tail of the celestial Scorpion. Locate the bright white star Theta Scorpii (magnitude +1.9). The same eyepiece ought to show you a faint (magnitude +5.3) companion. This is a bona fide binary system, located some 262 light years across the vacuous dark of interstellar space.

Just as the last rays of sunlight sink below the horizon, cloud-like structures appear, as if by magic, low in the eastern sky. Those ephemeral moments, on the cusp between twilight and true darkness, can bring out preternatural feelings. For soon, the full majesty of the Milky Way asserts itself, decorating the sky from northeast to southwest.

Grand Milky Way Sweep

Starting in Scorpius, pan your telescope, charged with a low-power, wide-angle eyepiece, slowly northeastward. Spend a minute or two exploring the field of view before moving on. Your telescope will carry you through mighty Sagittarius, Scutum and Aquila. Behold the vast shoals of stars of varying hues, the rich variety of nebulae, bright and faint congregations of stars—open clusters and globular

Fig. 15.3 The Summer Milky Way through Sagittarius (Image by the author)

clusters alike. Some loci are brimming over with starlight, but, every now and then, your gaze should chance upon regions that are utterly bereft of suns. Many a prominent (dust-laden) dark lane can be made out, veritable ink blots on the canvas of heaven. Next, move your telescope back to Scorpius, but by a different route; forget about method, just enjoy yourself.

Move your telescope east of Antares and north of the Scorpion's stinging tail. Use a star map to locate the beautiful color contrast pair **Omicron Ophiuchi.** Any small telescope will, at moderate powers, reveal to you the fifth magnitude primary shining in an ochre hue, and its magnitude +6.6 companion presenting as a blue–white sun some 10 arc seconds due north of its primary. Just south of Omicron lies the lovely triple-star system marked by 36 Ophiuchi. A small grab 'n' go 'scope ought to show you the beautifully matched A and B orange pair (both +5.2) separated by a mere 4.9 arc seconds of dark sky. The third component, curiously also orange, shines about 1.5 magnitudes fainter, about 12 arc minutes off to the east-northeast. Take your time drinking up the view; it's not all that often that you can frame three, fairly bright, marmalade stars in the same field!

Enough of doubles for the moment. Let's explore some treasured deep sky objects from the star fields of summer. First up, we'll take a visit to two globular clusters in Scorpius, **M4** and **M80**. Center sanguine Antares in the field of view of a low-power eyepiece. Next, push the bright star to the extreme eastern edge of the field. Now look at the western side of the field. A small, rich field telescope ought to allow you to see the sixth magnitude smudge representing M4 located there.

The first thing you'll probably note is that this globular cluster is not as bright as it first appears, at least based on its integrated magnitude. A high-power, wide-angle eyepiece will enable you to see this system a lot better and maybe even resolve some of the stars near its outer edges. A 4-in. telescope from a dark sky site ought to enable you to see a prominent north–south rift in the system. Can you see that, as globular clusters go, M4 is rather uncondensed? Can you see that it doesn't have a bright central concentration of stars? That's probably because M4 is one of the nearest of its kind to the Solar System—a mere 5,600 light years away, in fact.

Pan a few degrees in a north-northwesterly direction, and your finder 'scope ought to track down a fainter but much more condensed globular—**M80.** With an integrated magnitude of +7, it's considerably smaller and less conspicuous than M4, almost like the head of an approaching comet. You'll have a hard time resolving any stars in this globular cluster with small or moderate apertures. Why such a marked difference? Well, it's simply a question of distance. At 27,000 light years from the Solar System, M80 lies nearly five times further out than M4.

Our next port of call is the picturesque **Butterfly Cluster** (**M6**), which you can easily find about 5° north-northeast of the magnitude +1.6 star Lambda Scorpii, the constellation's second brightest luminary. From the UK, it often presents as no more than a hazy patch; at lower latitudes its true magnificence is made manifest. Though an easy binocular object, M6 looks far more impressive in a small tele-scope. A 3-in. 'scope ought to pull in a few dozen blue–white stars and a 6-in. up to 100 members. The eye is naturally drawn to the open cluster's brightest star, the lovely orange sun SAO 209132, situated at the cluster's eastern flank. M6 shines across 1,600 light years of interstellar space.

Next, we move to another jewel of the celestial Scorpion, **Ptolemy's Cluster** (**M7**), so named after the second-century astronomer, who recorded it as early as 130 BCE. Using his unaided eyes, Ptolemy likened this third magnitude glow to a "cloud following the sting of the Scorpion."

Lying about 4.7° east-northeast of the two bright stars that comprise the stinger (Upsilon and Lambda Scorpii), **M7** is the most southerly object in Messier's famous catalog. In a small grab 'n' go telescope equipped with a low-power, wide-angle eyepiece, the glittering stars of M7 sprawl over four full Moon diameters. Binoculars can show up several dozen members, while a 4-in. 'scope can pull in the best part of 100 stars in this magnificent open cluster. The system lies at roughly half the distance of M6 (950 light years).

Activity: Bonus points for spotting more open clusters in Scorpius. Start at M6 and draw an imaginary line from it to Zeta Scorpii to its southwest. Now pan your 'scope, equipped with a low-power eyepiece, slowly along this line. Did you manage to bag the fainter open clusters, NGC 6383, NGC 6281 and NGC 6242?

When we move from Scorpius into neighboring Sagittarius, the senses are heightened. We cast our gaze towards the center of the galaxy. And although obscured by starlight and dust, the epicenter of the Milky Way is marked by a radio source known as Sagittarius A, pinpointed to the celestial coordinates 17 h 46.1 m, −28° 51'. The eighteenth-century comet hunter Charles Messier had a field day in this constellation, bagging 15 new objects for his famous catalog.

First cast your gaze on the eight main stars making up the constellation. This is the naked-eye asterism known as the **Teapot**, although why this appellation was bestowed upon it is unclear. Nonetheless, we begin our tour here with a long, relaxed look at fifth magnitude M22, a beautifully rich globular cluster, far more impressive than any of its counterparts in the northern sky, located about midway between Mu and Delta Scorpii. Old Admiral Smyth described this system as "a fine globular cluster, outlying that astral stream, the Via Lactea, in the space between the Archer's head and bow, not far from the point of the winter solstice."

Spanning an area of sky about 80 % the area of the full Moon, a 3-in. telescope at moderate (~80×) will resolve a couple of dozen stars at its extremities. With a 4-in. glass, many more come into view. As you'd expect, increasing the aperture shows you more and more details inside this globular cluster. Would you agree that it looks less condensed than others of its kind? With a 6-in. 'scope and over, you can begin to discern that its brightest members are reddish in hue. M22 lies at a distance of 10,000 light years.

Return a low power eyepiece to your telescope and move it southwest until you center Lambda Sagittarii. About 1° to the northwest can you make out the much fainter (+6.8) globular cluster **M28**? At a distance of about 18,000 light years from the Solar System, you can readily understand why it appears so lackluster in comparison to mighty M22.

Next pan your telescope a few degrees further in a northwesterly direction with your low-power eyepiece. Soon, you'll chance upon the famous **Triffid Nebula** (**M20**), a prominent emission nebula. Don't be fooled by long-exposure photos, though. The nebula is far less impressive visually than through the eyes of a CCD camera. The vivid pink and blue colors garnered by the latter are not at all evident to those equipped with modest backyard 'scopes. A modest telescope ought to enable you to pick up some faint greenish hues, though. The Triffid lies at an estimated distance of 5,000 light years.

The spidery open cluster **M21**, shining with an integrated magnitude of +5.9, lies in the same field of view as M20, with a modest telescope revealing several tens of stars of the seventh magnitude sprawled across a plot of sky roughly one-third the size of the full Moon. But do bear in mind that this open cluster lies 4,000 light years inside the orbit of the Sun around the center of the galaxy.

For our next showpiece object in Sagittarius, center the star Lambda Sagittarii and move about 5.5° west. Here a low-power eyepiece should reveal the famous **Lagoon Nebula** (**M8**), a magnificent emission nebula spanning three Moon diameters, first observed back in 1680 by the British astronomer John Flamsteed (1646–1719). Moderate powers help maximize the contrast between the background sky and the nebula itself and should enable you to see a distinctive dark cleft that cuts the nebula in two. Indeed, a truly great view of M8 can be had on a mountaintop in California with a TeleVue 102 refractor equipped with a 9-mm hyper wide angle eyepiece. The eastern side of the nebula is bejewelled by a few dozen stars—comprising the open cluster NGC 6350—which renders the view quite unforgettable.

How do emission nebula glow?

The brightest star embedded within the nebula (nine Sagittarii) is hot and white, issuing forth a torrent of high energy ultraviolet light that, in turn, excites the gaseous matter within the nebula, causing to incandesce. Larger telescopes can reveal something of the so-called Hourglass Nebula, a prominent star-forming region brimming over with neonatal suns.

For our next object, we move north, close to the borders of Scutum, the celestial Shield. Locate Gamma Sagittarii and move your 'scope about 2.5° to the southwest. Here you'll find the celebrated **Horseshoe** or **Swan Nebula** (**M17**). It looks roughly wedge-shaped in a small telescope, but as the aperture increases, this bright emission nebula takes on an entirely new countenance. A 6-incher should allow you to see many prominent dark zones mottling what is otherwise a plainly incandescent green nebula.

Activity: Observers using large aperture 'scopes equipped with narrowband nebular filters have detected some of the many Bok globules embedded within M17. Find out what you can about these globules and their significance as 'smoking guns' for star formation.

M17 is thought to be one of the largest star-forming regions in the entire galaxy, easily dwarfing the Orion Molecular Complex explored earlier in the book. But unlike the Orion star-forming region adorning our winter skies, we view M17 face on. The complex is located 5,000 light years away.

Activity: Can you find **M18**, a rather inconspicuous star cluster located just 1° south of M17? Try a power of 100× to see if you can resolve the two dozen or so nineth magnitude stars within its borders.

Activity: If you don't have access to one yourself, try to borrow a good nebular filter from a club colleague. Go back and look at all the prominent emission nebula discussed so far—M20, M8 and M17. Threading the filter into your chosen eyepiece, compare and contrast the views with and without the filter. What conclusions can you rightly draw? Would the effects of the filter be more or less pronounced in an urban or rural setting? Why?

Moving north out of Sagittarius into the diminutive constellation of Scutum, the Shield, a small telescope equipped with a low-power eyepiece will present beautifully rich star fields to you. There are two open clusters well worth seeking out here. The first is the **Wild Duck Cluster** (**M11**), easily located as a sixth magnitude fuzz in binoculars a few degrees southeast of Beta Scuti. Located some 6,500 light years away, telescopes reveal up to 200 scattered over an area roughly half the size of the full Moon. The popular name of this cluster derives from its telescopic appearance only, which, with a little stretch of the imagination, resembles a flight of ducks. Larger telescopes begin to show that the core of this open cluster resembles a globular cluster, with some evidence of dark lanes emanating from the central regions.

For a lesson in contrasts, why not track down another open cluster within Scutum—**M26**. Binoculars or a finderscope shows it up as a seventh magnitude smudge just southeast of Delta Scuti. A small telescope will require high power (~100× or so) to even begin to resolve the individual stars. About two dozen can be resolved in a 3-in. telescope. M26 is actually closer than M11 at only 5,000 light years, yet it is obscured by interstellar dust that attenuates its true glory in comparison to the Wild Duck Cluster.

Activity: Observing the Great Rift: As well as containing myriad stars and other luminous bodies, such as emission and reflection nebulae, our galaxy is powdered with vast amounts of dust that, in places, can coalesce enough to almost completely obscure the light of background stars. Without these dark lanes, the Milky Way would look less majestic, less picturesque, as they add just the right measure of contrast with the bright star fields surrounding them, transforming the galaxy into something that is singularly beautiful to the human eye.

Late on June evenings at mid northern latitudes, you should be able to make out the three bright stars that mark the Summer Triangle—Deneb, Altair and Vega—a most convenient swathe of sky first proposed by the late Sir Patrick Moore. Center Deneb in a low power field of view and imagine a line running from this bright star to a spot roughly midway between the other two stars in the triangle—Vega and Altair. Run your telescope slowly along this line—the 'median' from Deneb if you like—looking out for chunks of sky that appear to have few or no stars. Such a sweep from a dark site will reveal to you the Great Rift of the northern Milky Way. Slowly, over the geologic ages, more and more stars will wink into view as these dust clouds dissipate.

Are there any regions of the electromagnetic spectrum available to us that can 'peer through' these visually opaque dust clouds?

The dust grains comprising these dark lanes are typically about 0.1 μm in size. And though their numbers are legion, astronomers are still at pains trying to understand exactly how they come to be. The cosmos is still rich in mystery.

Activity: Find out about cosmic dust, its composition, properties and theories of its formation. Does everything gel together (pun intended)?

Well, that ends our brief tour of the sky at high summer. I hope you will take the time to observe its great treasure of glories. Time waits for no one, though, so we must, as ever, move on to pastures new.

Chapter 16

Glories of the Fall

August and September are arguably the most joyous months for observers based in the British Isles, and many more countries besides. The return of dark skies after the twilight of summer, coupled with clement night time temperatures, conspire to create near ideal conditions for sky gazers—whether novice or veteran.

We begin our autumn vigil by exploring a particularly colorful theme—the red stars of early autumn—and the many delights they hold for those armed with modest backyard equipment. We begin with a real showpiece for small 'scope lovers, the beautiful color contrast binary pair **Alpha Herculis**. Known more affectionately as Ras Algethi, the third magnitude primary is an M-class red supergiant star with an emerald green secondary fully two magnitudes fainter separated from it by some 4.6 arc seconds of dark sky. These form a gravitationally bound system, with the secondary taking about 3,000 years to complete one orbit. It sure is an arresting sight in most any telescope.

From here, we move into Aquila, the celestial Eagle, where in the southern part of the constellation you'll find a seventh magnitude, blood-red carbon star **V Aquilae**, located about midway between Lambda and 12 Aquilae. It's a highly evolved sun, with a massive, sooty outer atmosphere that absorbs shorter wavelengths so effectively that scarcely any is emitted, explaining its intense, sanguine appearance. To the eye, the star's red color is obvious at a glance in the low-power field of a 15×70 binoculars but really comes alive in telescopes at moderate powers. Ultra-wide angle eyepieces and moderate magnifications frame this red giant using an 80-mm refractor amid the vastness of the Milky Way background stars.

From here, we move northwards again into Vulpecula, a small and faint constellation at the head of Cygnus. Once there, locate **Alpha Vulpeculae**—also known as Anser—a fairly inconspicuous fourth magnitude star that presents as a pretty binocular double with a less evolved orange giant star 8 Vulpeculae just 7 arc minutes

N. English, *Grab 'n' Go Astronomy*, The Patrick Moore Practical Astronomy Series, DOI 10.1007/978-1-4939-0826-4_16, © Springer Science+Business Media New York 2014

away. The connection between them is entirely illusory, however, with Anser being located only 297 light years distant but with its 'companion' lying nearly 200 light years further away. Some astronomers claim that Anser is in the process of brightening, perhaps as a consequence of the ignition of helium burning—which creates vast amounts of carbon—within its core. It will be interesting to monitor this system for signs of heightened activity in the coming years and decades.

Panning further north again into Lyra, the celestial Harp, and aiming our gaze towards the extreme southwest corner of the constellation, you'll soon pick up a magnitude +7.6 carbon star, **T Lyrae**. Despite its faintness, 10×50 binoculars show it very distinctly, owing to the relative paucity of hinterland stars in the field. This spectral class C8 is enormously luminous, however, shining with the brightness of several 100 suns. Only because of its enormous distance from us—an estimated 2,064 light years—does it appear so faint from our vantage.

Moving east into Cygnus, we locate a spot roughly $3°$ southwest of Sadr (Gamma Cygni), where binoculars will pick up the blood-red color of RS Cygni, shining feebly at magnitude +8. Located very close to the galactic equator, its brilliance is most definitely affected by obscuring dust (interstellar reddening) between our line of sight and this Mira type variable star, making it appear fainter than it really is. Careful study of this star reveals that its brightness varies by nearly two stellar magnitudes (7.2–9.0) over a period of approximately 400 days. While you're at it, why not move northeastward again towards Sadr to have a gander at the magnificent **Veil Nebula**—arguably the finest supernova remnant in all the heavens—which shows up nicely in a small refractor from a dark sky.

All the stars discussed thus far are bloated, highly evolved giants, but not all red stars are geriatric. To see one that bucks the trend, move your 'scope into the elegant constellation of Lacerta, spanning the sky between Cygnus and the most westerly stars of Andromeda. Located about $1.5°$ east of fourth magnitude star 11 Lacertae, 15×70 binoculars reveal an obviously red tenth magnitude star in the field of view called EV Lacertae. A 127-mm (5-in.) refractor puts up a marvelous show of this red dwarf star only 16.5 light years away. In days gone by, astronomers might have mistaken this M-class star as a 'late' or highly evolved sun, but we now know EV Lacerate is much younger than our own Sun and thus is spinning at a significantly faster rate. A rapid rotator usually generates very intense magnetic fields that twist and knot their way from the core towards the tenuous outer atmosphere—the corona—where they can ignite X-ray flares from time to time. And though astrobiologists are now considering such red dwarf stars as bona fide habitats for extraterrestrial life, the close proximity of their attendant habitable zones to these stars doesn't bode well for any emergent life forms they may harbor.

Continuing further north into the constellation of Cepheus, we wind down our survey of accessible red stars in the early autumn sky. Arguably the most famous carbon star of all, **Mu Cephei**, is easily located just a couple of degrees to the northwest of third magnitude Zeta Cephei. Lying some 2,800 light years from the Solar System, Mu Cephei enjoys the more famous appellation of the Garnet Star, so named by Sir William Herschel, who was greatly moved by its blood-red countenance. The Garnet Star is a red supergiant star and a prominent example of a so-called semi-regular variable, oscillating between magnitudes +3.4 to +5.1 and back over a period of approximately 2 years. Through an 80-mm refractor at 125×, Mu Cephei is almost addictive to look at, its cold, blood-red hue appearing like the distant embers of a campfire. You could stare at this marvelous sun for hours!

We end our sojourn with another carbon star within the borders of Cepheus, by paying a visit to the more elusive carbon star, **S Cephei**, located about midway between Kappa- and Rho Cephei. This star, made famous by studies conducted by pioneering American astronomer Henrietta Leavitt, based at Harvard College Observatory at the beginning of the twentieth century, this Mira-type variable exhibits an enormous change in brightness spanning nearly five visual magnitudes (+7.4 to +12.5) over a timescale of 486 days. Leavitt's work on Cepheid variable stars allowed other pioneers, particularly Edwin Hubble, to help establish the foundations of modern cosmology by establishing the enormous distances to the 'spiral nebulae.'

Exploring the Great Spiral in Andromeda

There are many objects in the heavens that augur the onset of autumn—the beautiful Pleiades cluster (M45) in Taurus, the great Square of Pegasus, or even the first appearance of blood-red Betelgeuse above the eastern horizon. But for many others, the darker nights of fall are inextricably linked to a fuzzy cocoon of light, easily found about 1° west of the magnitude +4.5 star, Nu Andromedae. We speak, of course, of the iconic **Andromeda Galaxy, M31** (NGC 224), a vast spiral galaxy, or 'island universe' as the astronomers of yore referred to it.

Probably known since antiquity, the first definite description of it goes back to AD 905. The Persian astronomer Al Sufi referred to it as a 'Little Cloud,' and thereafter it appears on many celestial maps dating from the pre-telescopic age. Simon Marius, a contemporary of Galileo, is usually credited with the first telescopic observation of M31 conducted sometime around 1612. And though telescopic equipment has vastly improved since this time, it is ironic that the basic description of Andromeda as a 'fuzzy cloud' still holds true today, even with large backyard instruments. The French comet hunter, Charles Messier, observed it on the night of August 3, 1764, when it was subsequently immortalized in his famous catalog.

Although early spectroscopic studies by Sir William Huggins revealed M31 to be largely stellar in nature as early as 1864, the true morphology of this object only became apparent when astronomers could make long exposure photographs of the object with large, observatory-class telescopes. Thanks to its modest 15° inclination from the edge-on position, the early astrographs were able to reveal a vast spiral galaxy, containing countless billions of stars, gas and dust, somewhat like a larger version of our Milky Way.

When astronomers figured out how to use standard candles in the 1920s—in this case, Cepheid variable stars of known intrinsic luminosity—they were finally able to pin down the vast distance to this galaxy, with the best modern estimates placing

Fig. 16.1 The majestic Andromeda Galaxy (M31) with its two bright satellites (Image © Mike Pearson. Used with permission)

it at 2.5 million light years away. Furthermore, earlier spectroscopic analysis carried out by Vesto M. Slipher carried out using the 24-in. refractor at Lowell Observatory, Arizona, in 1912, showed that the galaxy was blue shifted, indicating that it was hurtling toward us at an incredible 300 km per second. Today, we also know that the Andromeda Galaxy is the largest member—containing up to a trillion (10^{12}) suns—of the so-called Local Group of galaxies, consisting of more than 50 members, of which of our own Milky Way is a prominent component.

Although astrophotographers can produce stunning images of M31 with modest-sized telescopes from the comfort of their own backyards, there is something almost preternatural about letting your own eyes capture photons from this grand spiral galaxy (type SA), knowing full well that they first set out across intergalactic space when our early hominid ancestors were just beginning to walk upright across the African savannah.

Although best seen from a dark country sky, you can still make out M31 as a hazy spot in ordinary 7×50 binoculars, with an integrated magnitude of +3.4, even from the heart of a town or city. In small telescopes, such as an 80-mm (3.1-in.) refractor, the core is clearly discernible as a large, fuzzy oval that is considerably brighter toward the center, fading off to an indistinct edge on either side. Precisely how big the core looks actually depends on both the extent of light pollution in your local skies, as well as your telescope aperture. Even from the washed out skies of central London, for example, M31 looks for all the world like a large, unresolved globular cluster.

With a 4-in. (102-mm) 'scope in magnitude +3.5 skies, you might see an oval about 20 arc minutes across. But with a 10-in. (250-mm) 'scope, you'll make out something maybe twice as large from the same skies. And from darker locations, an 8-in. telescope or larger will begin to reveal the tell-tale signs of dust lanes that show up so beautifully in photographs. Modern ultra-wide eyepieces are especially useful for scanning Andromeda's vast dust lanes, as they simultaneously provide high magnification and wide fields of view. A 9-mm 100° eyepiece, for instance, is especially useful in this regard.

One valuable visual exercise is to see how far you can trace out the spiral arms extending either side of the bright core. With a 5-in. (127 mm) refractor and a power of about 125×, use averted vision to look for subtle attenuations in brightness as the faint outer arms give way to the dark of intergalactic space. Using this technique you should be able to follow M31's arms over 3° from edge to edge.

The Brighter Satellite Galaxies

M31 is known to harbor at least 14 satellite galaxies, most of which can only be resolved in very large telescopes. Luckily, though, it's fairly easy to pick out two of its prominent members. M32, a small elliptical galaxy, is M31's brightest satellite galaxy. Located as a magnitude +9.5 smudge, some 24 arc minutes south of M31's core, the light from M32 is much more condensed, possessing a very small but comparatively bright star-like core, helping M32 to be seen even with small 'scopes or medium-sized binoculars, even from heavily light-polluted skies. Higher magnifications, say 80× in a 80 mm 'scope, will help to bring out more details and confirm that it's definitely not a star.

The other prominent companion to M31, **M110** (NGC 205) can be seen 'hovering' a little over half a degree northwest of the great galaxy's core. Although considerably larger than M32 and displaying a brighter integrated magnitude at +8, M110 can actually be harder to spot owing to its much larger surface area (~22'× 11'). With a 5-in. (127-mm) refractor and employing moderately high powers of about 161× or so, you can begin to make out definite structure within this satellite galaxy. The nucleus, for example, is seen to be highly mottled inside a largely elliptical casing.

Fainter Still

More challenging again are **NGC 147** and **NGC 185** (also known as Caldwell 17 and 18) located some 7° to the north of M31. On a dark, transparent night, with the Moon tucked away, an 8-in. telescope ought to be large enough to bag them. Located just 1° apart, they can both be spotted in the same low power field using a wide-angle eyepiece. Shining feebly with integrated magnitudes of +9.5 and +9.2, respectively, these two dwarf galaxies, despite their physical association with Andromeda, are in fact located across the border in Cassiopeia! Discovered by John and William Herschel, respectively, a 5 in. refractor picks them up fairly well at 28× on the darkest nights with excellent transparency. With a concentrated gaze, you can see that they are morphologically quite distinct, with NGC 185 showing a more rotund morphology, while that of NGC 147 displaying a more bar-like shape. More magnification doesn't really help with these faint, extended elliptical galaxies.

Activity: A Globular Challenge

Returning once again to M31, over 500 globular clusters have been identified in its galactic halo. That's nearly twice as many as that of our own galaxy. By and large, you'll need at least a 10-in. reflector, a good high-resolution map, and a nice dark sky to make any inroads with these objects. With visual magnitudes of +13 to +15, high magnifications in the region of 200–400× are best for ferreting them out visually. Because of their tiny size, the vast majority of these globular clusters display little more than a stellar morphology with the exception of one, **Mayall II** (also called G1), which is viewable with a 12-in. Dobsonian.

Located about 3° southeast of the center of M31 (RA: 00:32:46.8/Dec: +39:34:42), this magnitude +13.8 object appears as a quasi stellar core with a fainter halo extending over just 5 arc seconds (about one-eighth of a Jupiter diameter). A widely spaced double star is located towards its southwestern edge.

Though very dim, G1 is a thrill to track down. Situated some 170,000 light years from the center of M31, it has the greatest absolute luminosity of any globular cluster known, being about twice as large as Omega Centauri in our own Milky Way. Very large backyard 'scopes, under the finest observing conditions, can track down a few other globulars embedded within the Andromeda Galaxy itself, but these will require high powers, large apertures and lots of patience to crack.

Well, there it is! The Great Spiral in Andromeda offers something for everyone's taste! From naked-eye visibility studies, through binoculars and then on through small and large telescopes, there's always something to test your observing prowess. Imagers, too, can have a field day with this extraordinary 'island universe' far across the sea of intergalactic space.

There are other galaxies worth visiting in the autumn sky. A short hop into Triangulum will show you another. Locate the third magnitude star, Alpha Trianguli, and pan about 4° west-northwest. Here, a low-power eyepiece will deliver a view of **M33, the Pinwheel Galaxy**. An 80-mm achromatic refractor picks up its brighter, more condensed, central regions, and averted vision will convince you that it's about twice as long as it is broad. With an integrated magnitude of +5.7, its light is spread out so much that it's quite difficult to see it with the naked eye. Clear, dark and transparent skies are a necessity with M33.

Fig. 16.2 The majestic Pinwheel Galaxy (M33) in Triangulum (Image courtesy of ESO)

A small telescope at low power will enable you to observe that the Pinwheel Galaxy presents a clockwise spiral. The many seventh magnitude stars 'surrounding it' seem like bees around a hive. These stars, of course, are quite unconnected with M33, located as it is some 2.3 million light years away. Indeed, the Pinwheel is widely believed to be a satellite galaxy of M31, which, in comparison, would swallow the former 15 times over.

There are a few other galaxies worth tracking down in the autumn sky with your grab 'n' go telescope. Point your telescope at Gamma Andromedae (Almach). Observe the beautiful color contrast binary system that it is. When you tire of looking at Almach (and you may not!), pan your telescope about 3.5° east. From a clear, dark sky, free of moonlight and dust, a 80-mm refractor equipped with a low-power eyepiece, picks up a linear array of faint suns between the sixth and the eighth magnitude. Once you locate this curious stellar arrangement, move your eye about a Moon diameter to the northwest, where the interesting edge-on spiral galaxy **NGC 891** (Caldwell 28) is situated. This is a challenging galaxy but ultimately very rewarding to locate and observe.

If you have difficulty seeing it, gently tap the outside of your 'scope. The eye sees it as just a smudgy tenth magnitude blob. Cranking up the magnification to 80× or more will show you the nebulous hub of this Sb type spiral. With averted vision and fully dark-adapted eyes, search for those spindle-like spiral arms projecting away from the central bulge. Larger telescopes, of course, will enable you to see NCG 891 more convincingly. A 5-in. refractor presents an utterly captivating scene at 169×, with a 12th magnitude star almost co-located with the central bulge. The very slender (~3 arc minutes wide) arms appear somewhat mottled and fade ever so gradually as they taper their way out from the galactic bulge. NGC 891 is one of the brighter members of the so-called Triangulum spur of galaxies, comprising some 40 members and located about 30 million light years from our Solar System.

Next we move to Pegasus and to the lovely island universe, **NGC 7331**, situated in the celestial equine's forelimbs, about 4° northwest of Eta Pegasi (Matar), a beautiful yellow star of the third magnitude, with a wide, blue companion of the nineth (a ghostly sight with the Oculus!) magnitude.

Back to NGC 7331. Shining slightly more intensely (magnitude +9.5) than the **NGC 891**, a low-power eyepiece reveals a faint, lens-shaped object slightly broader than the latter. As with nearly all distant galaxies, it benefits from higher power scrutiny. The eye is drawn to its comparatively brilliant, star like core, around which a diffuse fog of nebulous material can just be made out with averted vision.

The view is transformed in a 5- or 6-in. telescope, its spiral nature appearing quite obvious from a long, relaxed study in the dark.

Lying 47 million light years out, NGC 7731 is the brightest member of a small galactic clan, known prosaically as the NGC 7331 Group. The backyard telescope does not reveal a peculiarity that large, observatory-class instruments show. For some unknown reason, the bulge of this galaxy is rotating in the opposite direction to its spiral arms.

What might have caused this anomaly at the heart of NGC 7331?

September is usually a wonderful month for passionate deep sky observers. Mild nights and plenty of clear, stable skies provide a magical combination to lure sky gazers out of doors into the dark. High up in the sky at this time of year, to the east of the zenith, the capacious constellation of Cassiopeia is easily located by seeking out the characteristic 'bent W' appearance of its brighter stars. This is deep Milky Way frontier, snaking its way from northeast to southwest across the starry heavens. Our entire sky tour this month will be devoted to just a few of its legion treasures.

M34 is a fairly bright open cluster located just north-northwest of the midpoint between two celebrated stars—Almach, the famous double star in Andromeda, and Algo, the eclipsing binary star in Perseus. With the 40-mm Erfle delivering 19×, it is handsomely framed, showing a smattering of a dozen or so stars at first glance spread over an area roughly the size of the full Moon. But the real fun begins when you crank the magnification up to 83× with the 9-mm Oculus. Refocusing in delicious anticipation, the lovely double star H1123 appears at the cluster's epicenter. And sure enough, as you settle down to concentrate your gaze, the pair appears, pinpoints in plain white, shining with the eighth magnitude, surrounded by a few dozen young, 100 million year old suns, many fainter, but some significantly brighter than the binary jewels marking its epicenter. The system also contains the interesting close binary system, Struve 44, which has components of magnitudes 8.4 and +9.1. That'll be a good target for a larger telescope.

Is there anything suggestive about M34 that points to its relatively old cluster age?

Next we visit the heart of Perseus, visiting a few of the better known double stars that adorn the corpus of the Classical Hero. First up is **Epsilon Persei**, which presents as a tiny bluish spark of the eighth magnitude far away (possibly unconnected) from a third magnitude primary. Just 538 light years separates us from it. The even fainter companion to Zeta Persei is even more frustrating. But to be fair, the components here are even harder to disentangle. The primary shines proudly at magnitude +2.9, but the secondary is a paltry +9.5. Patience and a bit of experimentation with magnifications, however, pays off. Eta Persei, in comparison to the other Perseid binaries, is impressively easy. Like a fainter analog of Albireo, the beautiful soft orange light from the supergiant primary star contrasts with the faint, bluish secondary 29″ away. The 18× view was bonnie galore.

6 Trianguli is a more difficult target, owing to its faintness and relatively low altitude. That said, after letting the diminutive constellation rise a bit in the sky, you should be able to resolve this color contrast double with relative ease. The A/B separation is 4″, with a yellow G5 primary magnitude +5.2 and bluish secondary +6.7. It is well resolved at 150×. Astronomers know that this system, located a little over 300 light years away, is actually quadruple. It would be a lovely place for our descendants to visit (if they ever find a way of traversing the intimidating distances between the stars). They'll be no life on any of its worlds, most likely, but there might exist a cozy little place where humans, equipped only with a thin spacesuit, could enjoy some of the most spectacular sunsets in the galaxy.

By midnight on late September evenings, Aquarius is positioning itself nicely near the meridian and you can turn your small telescope towards the celebrated double star, **Zeta Aquarii**, the central star of the Water Jar asterism. Starting at 150×, the beautiful white pair are well resolved in an 80-mm glass; but the scene is even more impressive at 188×. The subtle differences between the two members become more apparent the longer you stare at them. And though the stars have a thin black line between them, it is very interesting to see the secondary as slightly more yellowed than its brighter counterpart, with a beautiful diffraction halo enveloping the entire system!

While you're in Aquarius, why not take a look at two Messier globular clusters, **M2** and **M72.** Located about 5° north of magnitude +2.9 beta Aquarii, M72 can be picked up in ordinary binoculars. In an 80-mm achromatic refractor at 83×, the magnitude +6.5 cluster presents as very compact but the stars remain unresolved. A 4-in. (102-mm) aperture is about the minimum to begin to resolve the individual stars in the cluster. Now move to M72, situated about 4° southeast of the magnitude +3.8 star Epsilon Aquarii. Mentally compare the views of M72 with M2.

Why is there such a great difference in brightness between the two globular clusters?

The answer, it seems, lies with the great difference in their distance from the Solar System. M2 lies only 37,000 light years away, while M72 is a staggering 56,000 light years away!

6 Trianguli was probably the most difficult target, owing to its faintness and relatively low altitude. That said, after letting the diminutive constellation rise a bit in the sky, you should able to resolve this color contrast double with relative ease. A/B separation 4″; yellow G5 primary magnitude +5.2 and bluish secondary +6.7. It is well resolved at 120× but more compelling to study at 200×. Astronomers know that this system, located a little over 300 light years away, is actually quadruple. It would be a lovely place for our descendants to visit (if they ever find a way of traversing the intimidating distances between the stars). They'll probably be no life on any of its worlds, but there might exist a cozy little place where humans, equipped only with a thin spacesuit, could enjoy some of the most spectacular sunsets in the galaxy.

Finally, when you get a good dark evening in late September and October, why not hunt down the nearest planetary nebula to the Sun. The **Helix Nebula** (NGC 7293), located some 700 light years away, can be found in a rather barren patch of sky roughly 20° due south of Zeta Aquarii. Due to its low altitude in most locations, it never presents as more than a round, greenish-gray glow, roughly half the size of the full Moon.

Dark Lanes in the Milky Way

The autumn sky is rich in stellar real estate, with the bright summer luminaries, Vega, Altair and Deneb—the celebrated summer triangle—still holding center stage. The great river of stars constituting the Milky Way runs across the sky from north to south, presenting endless rich pickings for those equipped with binoculars or a small telescope. It is the great contrast between dark and light that makes Milky Way scanning so addictive. Some dark patches seem empty to the naked eye with fainter stars only to be revealed by the truthful eye of the telescope. But there are numerous places where the telescope fails miserably, places that perfectly obscure light, like giant ink blots hanging between the stars. Despite its great faintness in even large amateur instruments, the winter sky holds perhaps the most famous of all, the Horsehead Nebula in Orion. Yet the autumn sky presents many more that are easy to see in binoculars or a small telescope.

These so-called dark nebulae were noted, cataloged and studied in great detail by the celebrated American astronomer Edward Emerson Barnard, together with Max Wolf, using the 40-in. Yerkes refractor at the end of the nineteenth century. Over a period of several years, they systematically cataloged 370 of these dark nebulae over vast swathes of the Milky Way. Subsequent research has revealed them to be cool, vast clouds that are opaque because of their unusually high concentration of internal dust grains.

The largest and most obvious of these structures is the **Great Rift**, a very dense cloud of space dust located about 300 light years away between Earth and the Sagittarius Arm of our galaxy and which runs lengthwise, through about one-third of its extent, from Cygnus in the north, right down to Scorpius in the south, flanked on both sides by lanes of faint background stars. Our journey begins in Cygnus using small, handheld 8×40 binoculars. First locate Deneb and draw an imaginary line from it to Sadr (Gamma Cygni). Moving your binoculars along this line, you're sure to notice a conspicuous dark region running along its eastern flank. You should see a markedly darker patch of sky, almost devoid of stars. This is the vast galactic tundra of gas and dust known as the **Northern Coal Sack.**

Cavernous Reaches

The Northern Coal Sack is a compelling sight, especially in a small telescope. What binoculars fail to uncover is a sense of the depth this cavernous region of interstellar space encompasses. Only at telescopic powers can one gain a real sense of how dense the dust cloud has to be to extinguish the background stars. Some fields of view serve up a patchwork of faint stars around kernels of impenetrable dark. The form of such dark clouds tends to be very irregular. Indeed, many have no clearly defined outer boundaries and sometimes take on convoluted serpentine shapes. Cranking up the power might allow you to glimpse extremely faint stars inside them. Such structures, astronomers inform us, came about over millions of years as the ash from myriad supernova explosions was dispersed in the winds of interstellar space. Doubtless, some has already been recycled into meteoroids, moons and worlds beyond our ken. But much of it, self evidently, litters vast tracks of our galaxy's spiral arms and in some locations marks the birthplaces of stars that are yet to come.

Our next port of call lies in the constellation of Aquila, where we'll visit a large and strange pair of dark nebulae known as **Barnard 142/143.** The structures are easy to locate in the sky just $3°$ northwest of brilliant Altair and $1°$ due west of Tarazed (Gamma Aquilae). Barnard 143, the northern void, which is about as large as the full Moon, as seen by the naked eye ($0.5°$), resembles a fork to some, with the prongs opening toward the west. Barnard 142 lies just to the south and is shaped like an irregular, starless rectangle. Both stand out quite nicely in binoculars, provided there is no moonlight or light pollution to spoil the view. Together, they remind many of a capital letter E (Barnard's E), also sometimes called the Triple Cave Nebula, It forms a well-defined dark area on a rich background of the Milky Way. Astronomers estimate that it lies at a distance of 2,000 light years from Earth.

Finally, we pan south into Scutum and to B112, a dark patch some 20×20 arc minutes in size located just south of the Wild Duck Cluster (M11). Again, this can be quite elusive if the sky is not clear and dark. Two other more fairly obvious dark nebulae in Scutum are **Barnard 111** and **Barnard 119a.** Both are found to the north-northeast of M11. Visually, they look like kidney bean-shaped patches that almost touch each other at their southern tips. Barnard 111 forms the northeastern edge of the Scutum Star Cloud and appears many times larger than Barnard 119a in photographs. Look for a "peninsula" of about half a dozen eighth- to tenth-magnitude stars wedged between the nebulae, as well as an L-shaped asterism, with a prominent orange central component just north of B119a.

These are just a few of the dark nebulae that litter the autumn sky. With patience and good sky transparency, you're sure to find many more. Good luck on your dark adventures.

High up in the sky at this time of year, to the east of the zenith, the capacious constellation of Cassiopeia is easily located by seeking out the characteristic 'bent W' appearance of its brighter stars. This is deep Milky Way frontier, snaking its way from northeast to southwest across the starry heavens. Our entire sky tour this month will be devoted to just a few of its legion treasures.

Our first target, **M52**, is easy to find. If you extend an imaginary line from Alpha to Beta Cassiopieae and continue on about the same distance again, binoculars will show a small fuzzy patch nestled in a rich milieu of background Milky Way stars. An 80-mm (3.1-in.) glass at 33× reveals a sprinkling of seventh magnitude stars arranged in a vague, fan shape, and extending over an area roughly one-third the size of the full Moon. The view is better at 100×, where several dozen stars can be resolved. But if you turn a big Dob, of say 30 cm or larger on this cluster, you'll be delighted with a magnificent view of over 100 stars, varying in brightness by at least two magnitudes.

While you're in the vicinity of M52, why not have a go tracking down the elusive **Bubble Nebula** (NGC7635), just over half a degree to the southwest. You'll need a 10-in. and dark skies to make any inroads on this object, the brightest section of which surrounds an 8.5 magnitude star—the energetic OB class HD 220057, whose stellar winds have blown away gas near the star and carved out a bubble of ionized glowing gas. The star HD 220057 is a hot massive giant some 40 times more massive than our Sun. I'd take pity on anyone who gets too close to HD 220057. The torrent of high energy radiation issuing from this stellar hothouse will turn you to toast in minutes!

Our next object is often overlooked but is truly one of those targets that looks good in almost any instrument. The magnificent sprawling open cluster **NGC 7789** is found about midway between Sigma and Rho Cassiopeiae, showing up as a fuzzy patch about half the size of the full Moon in 15×70 binoculars. A 4-in. (10.2-cm) refractor at 37× reveals 40 or 50 stars arranged loosely across the telescopic field. But if you turn a 10-in. 'scope on it, the view is transformed into over a hundred members. The more you look at NGC 7789, the more shapes you see. What forms do you make out?

Our next target, **NGC 185**, presents something entirely different. To see it well, you'll need at least a 4-in. instrument and dark, transparent skies, though it can be picked up in moderately light-polluted skies with larger apertures. This dwarf galaxy lies about half way between Alpha Cassiopeiae and the Andromeda Galaxy (M31). Shining at magnitude +10, it presents as a small, 12×10 arc minute oval of fuzzy luminosity at moderate high magnifications (100×). NGC 185 is a member of the Local Group and a satellite galaxy of M31. While you're at it, why not seek out its near twin, NGC 147, located less than 1° west of NGC 185.

We return to the opulent confines of the Milky Way at the heart of Cassiopeia by taking a visit to a bright galactic star cluster, **NGC 457**, located about 4° to the southeast of Gamma Cassiopeiae. Many appellations have been bestowed on this cluster, but perhaps the most fitting—at least in small 'scopes—is the ET Cluster, on account of its resemblance to Steven Spielberg's anthropomorphic alien. It's a fine sight in a 4-in. refractor at 100×, which reveals a sprinkling of a several dozen white stars dominated by the brightest member, Phi Cassiopeiae. Larger apertures are especially useful at fleshing out the magnitude +8.6 red giant star at the heart of the cluster, its distinctive garnet hue really shining through with telescopes of 5 in. (12.7 cm) or over.

We're going to round up our sojourn through the Heavenly Queen by scrutinizing some individual stars. Our first port of call is the lovely sun-like star **Eta Cassiopeiae.** Shining at magnitude +3.5, it gives us a glimpse of what our Sun would look like at a distance of 19 light years. In a 3-in. glass at 75×, Eta is shown to have a fainter, magnitude +7.5 companion about 13 arc seconds away to the northwest, which to some has a rich red color. In a larger instrument, such as a 5-in. (127-mm) refractor, the color contrast is especially striking. The system forms a true binary pair with an orbital period of about 480 years. The companion will continue to widen by a few more arc seconds until 2050.

Extending a line starting at Delta through Epsilon and continuing on northwesterly the same distance again you'll soon chance upon the easily overlooked **Iota Cassiopeiae.** At low power in a small telescope, this magnitude +4.5 whitish star looks nothing out of the ordinary, but push the magnification up to 150× or higher in even a 60-mm (2.4-in.) 'scope and you'll see that it has a magnitude +8.4 companion about 7 arc seconds away to the southeast. But there's more still. The system is actually a triple, with another, more difficult seventh magnitude component closer in, separated from the primary by a sliver of dark sky just 2.7 arc seconds wide. All three are a real challenge for a 60-mm scope, but a 3-in. f/9 glass bags all of them on a good night.

Our final port of call on this whistle stop tour of Cassiopiea is arguably one of the most accessible and well-studied variable stars in the northern sky—**RZ Cassiopeiae.** Although not the brightest eclipsing binary in the sky, RZ Cassiopieae (Cas) is one of the easiest to observe visually. Its magnitude range is 6.18–7.72, and the minima last for only 4.8 h. That is quite a dramatic change and is easily monitored in binoculars or a small backyard 'scope. Indeed, for newcomers to variable star observing, RZ Cas still offers the possibility of estimating stellar magnitudes, allowing you to record clear variations and getting feedback in the form of a light curve, all within a single night!

That ends our selective tour of the autumn sky. There are, of course, many other objects that can be visited. By now, though, you'll have learned a great deal about many of the showpiece objects in the sky, their varied nature and significance in the scheme of things. The most important thing to remember is that visual astronomy is a labor of love and gets more addictive the more frequently you observe.

Glossary

achromat A telescope that uses traditional crown and flint glasses to bring two colors of light to a sharp focus.

apochromat A telescope that makes use of exotic materials such as fluorite and synthetic low dispersion glass to bring three colors of light to a sharp focus.

barycenter The point that is coincident with the center of mass of a system of gravitationally interacting bodies.

beta Lyrae stars A group of binary stars that vary in brightness owing to one component eclipsing the other.

Big Bang The explosive event that brought the universe—its matter and energy—into existence some 13.8 billion years ago.

black hole A region of space where the gravitational field is so intense that light itself cannot escape from it.

Catadioptric A type of compound telescope employing lenses and mirrors to collect and focus light. Catadioptrics (CATs) include Schmidt and Maksutov Cassegrain telescopes.

Cepheid variable A class of evolved high mass stars that vary in brightness as they pulsate.

chromatic aberration This is the unfocused light seen in achromatic telescopes that causes a purplish tinge around bright objects.

comet Icy body made from the pristine volatile substances that formed in the early Solar System.

Dawes limit An empirical formula created by William Rutter Dawes in the nineteenth century, it states that the resolution limit (in arc seconds) of two equally bright stars of the sixth magnitude is given by $4.57''/D$, where D is the aperture of the telescope in inches.

degeneracy pressure This is a kind of pressure exerted by subatomic particles such as electrons, which are compelled to obey the rules of quantum mechanics.

exit pupil This is the size of the light cone that enters the eye via the eyepiece. It can be calculated by dividing the telescope aperture (in mm) by the eyepiece magnification.

filter A device that transmits selected wavelengths of light, which can be employed visually or photographically.

focal length The distance from the main lens/mirror of a telescope to its focus.

galaxy A vast collection of stars, gas, dust and dark matter, held together by mutual gravitational forces. A typical galaxy contains 100 million stars and are observed in spiral, elliptical and irregularly shaped forms.

globular cluster A tightly bound system of very ancient stars (usually containing between 100,000 and 1 million members) found mostly in the halo of our galaxy and other galaxies.

light year The distance that light travels in a year.

magnification This is calculated by dividing the focal length of the telescope by the focal length of the eyepiece.

multi-coated Multiple layers of anti-reflection coatings applied to the surface of a lens.

neutron star The super high density core of a star that has ended its life as a supernova. They are just a few tens of kilometers across and are composed mostly of neutrons.

open cluster A gravitationally bound system of stars usually formed from the same cloud of gas and dust.

parabolic mirror This refers to the shape of the concave mirror used in modern Newtonians. Only paraboloidal mirrors are fully corrected for spherical aberration.

planetary nebula A vast shell of gas and dust released by a sun-like star at the end of its life. Short wave radiation from the core of the star causes the material to light up like a bulb, producing their pretty colors.

standard candle An object whose properties are sufficiently well known that its absolute luminosity varies little between similar examples of the same object. Cepheid variable stars are one such object. Cosmologists use standard candles to establish intergalactic distances.

true field of view This is calculated by dividing the apparent field of view of the eyepiece by its magnification.

white dwarf The burned out core of a sun-like star, usually composed of carbon nuclei and about the size of Earth.

References and Bibliography

Star Atlases and Constellation Guides

J. Mullaney, W. Tirion, *The Cambridge Double Star Atlas* (Cambridge University Press, Cambridge, 2009)

P. Simpson, *Guidebook to the Constellations* (Springer, New York, 2012)

R.W. Sinnot, *Sky & Telescope's Pocket Sky Atlas* (Sky Publishing, Cambridge, 2006)

W. Tirion, *The Cambridge Star Atlas* (Cambridge University Press, Cambridge, 2011)

Introductory Visual Astronomy Texts

G. Consolmagno, D.M. Davis, *Turn Left at Orion* (Cambridge University Press, Cambridge, 2011)

S.J. O'Meara, *The Messier Objects* (Cambridge University Press, Cambridge, 1998)

F.W. Price, *The Planet Observer's Handbook* (Cambridge University Press, Cambridge, 1994)

I. Ridpath, W. Tirion, *Stars & Planets* (Princeton University Press, Princeton, 2008)

Gear-Orientated Texts

T. Dickinson, A. Dyer, *The Backyard's Astronomer's Guide* (Firefly Books Ltd, Richmond Hill, 2008)

N. English, *Choosing and Using a Refracting Telescope* (Springer, New York, 2011a)

N. English, *Choosing and Using a Dobsonian Telescope* (Springer, New York, 2011b)

N. English, *Grab 'n' Go Astronomy*, The Patrick Moore Practical Astronomy Series, 241
DOI 10.1007/978-1-4939-0826-4, © Springer Science+Business Media New York 2014

Astronomy Magazines

Astronomy. www.astronomy.com
Astronomy Now. www.Astronomynow.com
Sky & Telescope. www.skyandtelescope.com

Index

N. English, *Grab 'n' Go Astronomy*, The Patrick Moore Practical Astronomy Series,   243
DOI 10.1007/978-1-4939-0826-4, © Springer Science+Business Media New York 2014